Amphiprion frenatus *(Red Clownfish)*

Other titles in this series:

The Tropical Aquarium
Community Fishes
The Healthy Aquarium
Aquarium Plants
Fish Breeding
Koi
Fancy Goldfishes
Livebearing Fishes
Central American Cichlids
African Cichlids
African and Asian Catfishes
South American Catfishes

A FISHKEEPER'S GUIDE TO

MARINE FISHES

Zanclus cornutus *(Moorish Idols)*

Chelmon rostratus *(Copper-banded Butterflyfish)*

A FISHKEEPER'S GUIDE TO

MARINE FISHES

A superbly illustrated introduction to keeping marine tropical fishes, featuring over 50 magnificent species

Dick Mills

Tetra🐟Press

No. 16064

A Salamander Book

Synchiropus picturatus *(Psychedelic fish)*

Credits

Editor: Geoff Rogers   Designer: Tony Dominy
Colour reproductions:
Rodney Howe Ltd.
Filmset: SX Composing Ltd.
Printed by Proost International Book Production, Turnhout, Belgium.

Author

The author, Dick Mills, has been keeping fishes for over 30 years, during which time he has written many articles for aquatic hobby magazines as well as 12 books. A member of his local aquarist society, for the past 20 years he has also been a Council member of the Federation of British Aquatic Societies, for which he regularly lectures and produces a quarterly News Bulletin. By profession, he composes electronic music and special sound sequences for television and radio programmes – a complete contrast to fishkeeping, the quietest of hobbies.

Consultant
Fascinated by fishkeeping from early childhood, Dr. Neville Carrington devised an internationally known liquid food for young fishes while studying for a pharmacy degree. After obtaining his Doctorate in Pharmaceutical Engineering Science and a period in industry, Dr. Carrington now pursues his life-long interest in developing equipment and chemical products for the aquarium world.

Contents

Introduction

A few years ago, almost the only places you could visit to see marine fishes were public aquariums; people fortunate enough to keep marine fishes successfully in their own homes were very few and far between.

In recent times, thanks to television, the colourful fishes from the coral reefs of the world have almost literally swum into our homes to beguile us with their beauty. This continual exposure to the delights of tropical underwater life resulted in an upsurge of interest in keeping marine fishes in the home aquarium. Initially the fishes were almost prohibitively expensive due to the high freight charges coupled with the fairly low proportion of fishes that arrived safe and well. It was not surprising that the lifespan/ expectancy of the majority of aquarium-kept marine fishes was reckoned in months rather than years. Such was the price that had to be paid by the early

pioneers in their quest for practical experience.

However, all that changed considerably with the introduction of jet air transport and expanded polystyrene. The former cut the traumatic journey times quite drastically; the latter reduced the equally damaging temperature losses to a minimum; and both helped enormously in ensuring that a much higher proportion of fishes arrived at their destination in a good state of health. The main reason for fish losses in transit today is the delay that occurs when fish and paperwork arrivals do not coincide.

Healthier fishes from the outset, together with the knowledge gained from experience and the increase in quality of foods and equipment necessary to maintain marine fishes, mean that the marine fishkeeper today is in a far better position to get off to a confident start in what is often seen as the ultimate in fishkeeping.

The attraction of marine fishkeeping

The answer to the question "Why keep marine fishes?" is not hard to find; the fishes are so beautiful that they positively cry out to be kept, particularly in a home aquarium where you can see them at your convenience and in comfort whenever you wish. But there are other reasons too, especially if you are a fishkeeper already.

You may have started keeping fishes at the tiddler level and then moved on to tropical or coldwater freshwater species as your interest grew. You may see keeping marine fishes as a continuing challenge to your fishkeeping skills, a different avenue to explore. There is also no little pride in being able to keep marine fishes successfully in the home aquarium, although whether you will ever feel able to emulate other fishkeepers in displaying your fishes at exhibitions is another matter altogether! On the other hand, you may want to keep marine fishes simply because, as far as you are concerned, there will never be anything that appeals to you more – which almost brings us back to the beginning again.

Think hard before you start. Marine fishkeeping is a much more complex and technical subject than freshwater fishkeeping. Although it is not strictly

necessary, a practical grounding in basic fishkeeping will be beneficial, because much of the experience gained previously will save you time, worry and maybe expense as you concentrate on the extra important differences of saltwater aquarium management. Because all marine fishes are imported wild-caught species, not bred in commercial numbers as are their tropical freshwater counterparts, prices of individual fishes are proportionately higher; but you will not be keeping anything like the same number of fishes in your aquarium.

One of the fascinations of keeping marine fishes is that, because it is such a relatively new aspect of the hobby, there is still a great deal to find out about it, particularly in the breeding of the fishes. With a lot of dedication and hard work (plus a generous slice of good fortune) there is still a chance for you to become one of the pioneers of modern marine fishkeeping.

The marine fish
If you disregard for a moment any comparison between the colours of freshwater and saltwater fishes, you will need to appreciate that there are other less apparent differences between the two types.

The main difference that the marine fish has to cope with is being surrounded by salt water rather than fresh water. If we accept that both types of fish have body fluids of the

Below: *The beautifully coloured Royal Gramma or Fairy Basslet* (Gramma loreto) *finds plenty of sanctuaries in this superbly decorated aquarium.*

FRESHWATER FISH

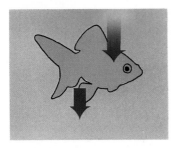

MARINE FISH

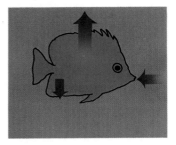

Above: *A freshwater fish (left) constantly absorbs water into its body by the natural process of osmosis. A marine fish (right) loses water by the same process and must drink heavily to compensate.*

same strength or concentration, it will make understanding the problem simpler. The freshwater fish is surrounded by a weaker strength fluid (fresh water), but the marine fish is surrounded by a stronger fluid (seawater). Due to *osmosis*, there is a natural tendency for water to move from a weak solution towards a stronger one when separated by a semi-permeable membrane, in this case the fish's skin and gills.

Applying this to the two fishes in turn, you can see that the freshwater fish is always absorbing water through its skin, but the marine fish is always losing it. In aquarium terms, the freshwater fish needs to excrete large quantities of water (mostly as urine) to prevent itself from being 'drowned', whereas a marine fish has to drink a lot of water to replenish continuous losses through the skin, excreting very little water but much salt and minerals.

The majority of tropical freshwater fishes' habitats are subject to fluctuating conditions throughout the year, especially when monsoon rains occur. Through evolution, the freshwater fish has developed the means of coping with these environmental changes, and some can actually make the transition from fresh to salt water and back again quite successfully.

The saltwater fishes come from a much more stable environment (97%

of the water covering the earth is salty), and have therefore had no need to adapt themselves to changes in their environment. You must provide an aquarium large enough not only, as in freshwater aquariums, to give adequate water surface area, but also to ensure that the water conditions remain as stable as possible. Marine fishes are very susceptible to rapid changes in the aquarium conditions.

Another factor to be considered is that some marine fishes have a symbiotic relationship with other fishes or invertebrates. Two examples of this are the Cleanerfishes, which service other fishes by removing parasites from their skin and gills, and the sea anemones, which provide a home for the Anemonefishes.

Choosing fishes
The choice of fishes for the aquarium is very wide, though they are more expensive than fishes for the freshwater tropical aquarium; but you must be careful when stocking the aquarium. Fishes from the coral reef are used to well-oxygenated, well-lit waters. Those fishes requiring small plankton-sized foods will not thrive in a clinically clean aquarium. Most fishes are territorial, too, and in the confines of the aquarium, where they cannot swim away from trouble (or maybe just find some peace and quiet), it is advisable to keep only one of each species – which rather puts a damper on any breeding plans!

Correct stable conditions and suitable tankmates are among the most important factors to consider when planning a marine aquarium.

The marine environment

Faced with the prospect of providing an environment for marine fishes, most people would try to simulate in the aquarium a close approximation to the conditions that the fishes are used to in nature, with no technical aids at all except for the heating apparatus and maybe some aeration. This is a natural thought process, but there are some immediate drawbacks, not the least of which is where to get the sea water from? Another problem is that if anything upsets the delicate balance of the water conditions (some decaying uneaten food, or an unnoticed dead fish) the whole aquarium will become polluted in a very short space of time.

An alternative method is to set up the aquarium so that you have *total* control over the temperature, lighting and water conditions. In this situation, the aquarium positively bristles with technological aids such as protein skimmers, ozonizers and ultra-violet sterilizers, and woe betide any speck of dirt or algae that dares appear on the pure white coral!

An in-between path tries to provide the natural requirements of the fishes but uses modern technology in the process. Of the three methods, this is probably the one practised by most marine fishkeepers, for two simple reasons – it seems to work better than the first of the methods described earlier, and it is less expensive than the second.

For convenience and reliability, synthetic sea water is used, together with much of the technology found in freshwater fishkeeping. Experienced fishkeepers, coming from other areas of fishkeeping, will neither have to forget all their old rules nor have to learn totally new ones. Apart from a couple of items of equipment (neither of which is absolutely mandatory), the cost of the hardware is exactly comparable to that of the freshwater tropical aquarium; the only extra outlay will be the cost of the synthetic salt mix. Decorative material costs balance out much the same; the high price of the coral sand and corals is offset by there being no plants to buy.

This semi-natural method, as it has become known, is the method that we shall follow in this Guide.

Which animals to keep?

There is a wide range of marine animals that can be kept under aquarium conditions; in addition to the fishes themselves, there are the invertebrates, some of which are very necessary to keep with certain fishes.

On the other hand, not all marine fishes are of tropical origin. Fishkeepers living in temperate countries will find much of aquatic interest in rockpools left by the receding tide around the coast. A small aquarium can provide a home for many of the rockpool's

Above: *Every marine fishkeeper's idea of heaven – diving among the fishes over a tropical coral reef.*

inhabitants; and of course, if the coast is conveniently near, the cost of replacement stock (or even sea water itself) is limited to the price of the transport used. Nearness to the coast, however, should not be an excuse for using local sea water in the tropical marine aquarium.

Sea water must be regarded as a living entity, and many of the organisms in the cooler sea water will not survive under tropical temperatures, thus providing an immediate pollution hazard. Again, sea water must be clean, and although it might suffice for short-term purposes, such as a collection of native marine rockpool fishes and invertebrates, it is unlikely that water collected from a convenient beach will be clean enough for long-term, large-scale aquarium use.

Below: *No fishes in this aquarium, but a collection of invertebrates can be just as fascinating to contemplate.*

The tank

Two things are vitally important when selecting a tank for marine fishes; it must be large enough, and it must be completely impervious to the highly corrosive salt water and spray. Once you have satisfied yourself on these two factors then you can consider further incidentals such as the style, and where you are going to site it.

Selecting the correct size

Experienced freshwater fishkeepers will already be familiar with the theory that the number of fishes that can be comfortably held in a tank of any given volume depends on three things: the amount of dissolved oxygen in the water and its replenishment rate capability, the temperature of the water, and what sort of fishes you are keeping. Warm water holds less dissolved oxygen than cool water, but coldwater fish require more oxygen than tropical species. The whole complex problem can be calculated in a convenient manner by allowing a certain total of fish body lengths (measured in centimetres or inches) per given area (measured in corresponding square centimetres or square inches) of water surface.

Below: *The fish-holding capacity of a tank can be estimated in relation to the area of the filter base or water surface i.e. approximately 1 cm of fish per 120cm^2 of water surface.*

This theory is not entirely transferable to the marine aquarium, for many of the fishes we intend to keep are used to coral-reef water, which may be over-saturated with oxygen. However, there is another way of calculating the fish-carrying capacity of the aquarium, based on the rate at which bacteria in an undergravel biological filtration system can safely purify the aquarium. (This process is explained more fully in a later chapter.) This means that every centimetre of fish body length requires a certain area of filter bed surface. Because the undergravel filter is usually fitted to cover the whole base area of the aquarium, you can conveniently regard the filter bed surface area in the same way as the water surface area. As a starting guide, allow approximately 120cm^2 of water surface area for every 1cm of fish body length. (Using Imperial units this corresponds to 1ft^2 of water surface area for every 3in of fish). This formula must only be used for aquariums in which full base biological filtration is used; it will produce incorrect results (and dangerous overcrowding) if used in aquariums with conventional external filters.

You also have to take into account that the marine fish comes from a very stable environment. In aquarium terms, the larger the volume of water the more stable its conditions will

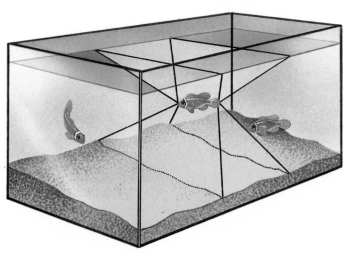

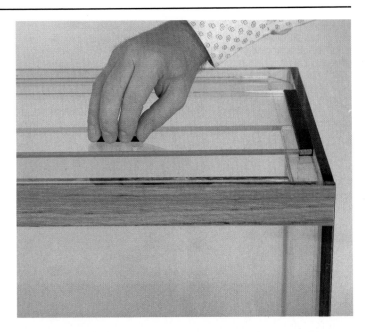

Above: *Saltwater spray can be very damaging, particularly to light fittings. This snug-fitting, sliding cover glass* *prevents splashes, cuts down excessive evaporation and makes feeding the fishes a lot easier too.*

remain. A large tank will also provide ample swimming and hiding space for the fishes, which can be very territorially minded, often to the point of being positively anti-social.

Bearing all this in mind, it is clear that the tank should be fairly large for several reasons, not least of which is to ensure that you have enough fishes to look at! It is recommended that you should not attempt to keep marine fishes in any tank measuring less than 900mm × 300mm × 380mm (36in × 12in × 15in), and holding approximately 104 litres (23 gallons) of water when completely set up.

Tank construction
Tanks made from any material that is likely to be affected by the corrosive action of salt water are obviously unsuitable for marine fishes. While some plastic- or nylon-coated metal-framed tanks may appear to be adequately protected, these generally last only a short time and, as they are in any case no longer readily available, you would be well advised to choose a suitable

tank in the first place, and avoid building up problems for yourself and your fishes later. Metal reflectors and hoods are also vulnerable to attack and any condensation dripping back into the tank could be toxic to fishes.

Silicone-sealed, bonded all-glass tanks are eminently suitable as saltwater aquariums, being completely impervious to saltwater damage. They have the added advantage that, because of the relative ease of construction, they can be made to fit any vacant site planned for the aquarium, or of almost any complex shape that you care to imagine. Make sure, with any large all-glass tank, that there are adequate bracing supports across the top of the tank to prevent the front glass from bowing outwards under the water pressure.

Marine aquariums can be made from any inert material; wood, concrete, or fibre-glass can be used quite successfully as basic structures, with a glass viewing panel bonded into the front. The inside of these tanks (do not forget the inside of the matching wooden reflector hood too!) must be suitably sealed with epoxy or polyester resins, and cured with strong sea water before use. For very large aquariums, wooden construction designs offer a relatively cheap method of tank manufacture.

One advantage of making tanks to your own specification, particularly with fibre-glass, is that you can incorporate filter systems into the design and they will be completely undetectable to the front-of-aquarium viewer.

Siting the tank

Despite the fact that coral fishes are used to bright light for most of the day, it is wrong to place the tank where it receives a large amount of sunlight. A window location results in over-heated water in summer and over-cooling in winter. Unlike freshwater aquariums, however, the growth of beneficial algae is to be encouraged, and a window location cannot be dismissed on this account.

A position that receives little or no direct daylight is ideal; it is far easier to provide the necessary light artificially than to block out or otherwise control

unwanted excess natural light.

Take care to choose a site capable of supporting the aquarium's considerable weight; our suggested 900mm (36in) tank will weigh at least 150kg (330lb) when fully furnished. A firm, level support is essential for all aquariums; stand all-glass tanks on a slab of expanded polystyrene to cushion them against any irregularities in the supporting surface, which might otherwise set up dangerous glass-cracking stresses. Very large tanks must be supported by strong braces especially built into the house structure itself.

The aquarium should also be near an electrical power outlet (for the aquarium's heating, lighting and filtration equipment) and there must be easy access for feeding and maintenance. Finally, make sure that there is room for your favourite armchair nearby, to complete your fishwatching comfort!

Below: *Who could carry on watching television and ignore the spectacle of this magnificent marine aquarium? It is the product of a wealth of practical experience, artistic appreciation and creative ability of the highest order.*

Heating and lighting

No heating will be required for marine fishes collected from temperate beaches and rockpools; they may need cooling in over-warm room conditions. However, the colourful coral reef fishes (the main attraction of marine fishkeeping) will need the water temperature of their aquarium to be kept fairly constant at around the 24°C (75°F) mark, similar to the water temperature of their natural home in the tropical seas.

There is no need to worry if the temperature varies a degree or two on either side of that pre-set by the thermostat; the fishes have a reasonable tolerance to the overall temperature range, and most are able to survive quite moderate extremes as long as these upper or lower limits are arrived at gradually. As you will learn in fishkeeping, the rate of any change of environmental conditions is more critical than the actual values.

Provision of heat for the aquarium water is simplicity itself, as thermostatically controlled miniature immersion heaters are widely available in aquatic stores. Measurement of aquarium water temperature is done by means of an inexpensive thermometer, which may be internal or external. It is wise to invest in a mercury or spirit-filled internal thermometer for reliability and accuracy. External liquid crystal types are often affected by room temperatures and are relatively shortlived.

Types of heating equipment
When heated individually, all modern aquariums are heated by electricity using submersible heating coils mounted in watertight tubular glass containers. In order to maintain the water temperature within narrow limits, the supply of electricity to the heating coil is continually switched on and off by a thermostatic device set to the required average temperature.

For the fishkeeper with more than one aquarium, and who may well have many aquariums in one room or fish-house, it may be financially advantageous to heat the whole room-space rather than each individual tank. This can be easily

arranged with a thermostatically controlled radiator or fan heater; even a carefully adjusted paraffin (kerosene) heater will be effective, but a certain amount of heat-saving insulation must be incorporated in any space-heated room to keep heating costs to a minimum. In this Guide we shall concentrate on individual aquarium heating.

Immersible combined heater-thermostat units are now common. Modern thermostats have solid-state circuitry (more accurate than outdated electro-mechanical bi-metallic strips). Similar thermostats requiring separate immersible heaters may be mounted remotely from the tank provided that the temperature-sensing probe is submerged in the water.

When choosing heating equipment for the marine aquarium, be careful to select only that which has been made using saltwater-tolerant materials; and this goes for any associated equipment too. Rubber will perish

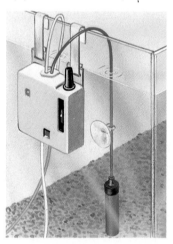

Above: *This external thermostat has microchip circuitry and a submerged sensor for precise heat control.*

Outer glass casing

quickly, and any metal must not come in contact with the water and is best excluded totally from the marine aquarium. If you are using a separate external thermostat, this is generally hung on the aquarium by means of a metal clip or strap: the metal wire springclip must be covered with a length of plastic airline (tubing) before being attached to the thermostat body, to prevent it coming in contact with the aquarium water. Covering the metal strap fitted to the solid-state type of thermostat, and usually made of stainless steel, might be more problematical; a solution might be to stick the thermostat to the outside of the aquarium with a blob of silicone sealant. Clips used to hold the heater in position in the aquarium must be of non-toxic materials (in freshwater aquariums this is just as vital), and plastic materials are far safer than metal, especially in contact with highly corrosive salt water.

Size of heaters
Once the aquarium water has been heated to the desired temperature, it requires relatively very little electricity to maintain it at that level, provided that the aquarium is large enough to keep a stable reserve of heat. With the suggested minimum size aquarium of 900mm × 300mm × 380mm (36in × 12in × 15in) containing about 104 litres (23 gallons), and allowing 2 watts of heat per litre (or 10 watts per gallon) you will need a heater with an output of 200 watts for

use in a normal living room.

Aquarium heaters are made in standard sizes (wattages) and, although there may well be a single heater of exactly the correct size, for an aquarium larger than 600mm (24in) long it is advisable to split the total heat requirement between two heaters (say of 100 or 150 watts each), so that the heat is distributed more evenly throughout the aquarium. This will mean buying two combined heater-thermostat units if this is the method you intend to use. (This will provide a degree of insurance against component failure, as well.) Where the heat is to be controlled by a separate external thermostat, you can operate two separate heaters by one thermostat, but make sure the electrical current-carrying capability of the thermostat exceeds that required by the heaters.

In either event, place the heaters at opposite ends of the aquarium. Combined units are usually placed semi-vertically (or vertically, if the depth of water allows); single heaters may be mounted horizontally, but clear of the gravel, in clips fixed to the glass by non-toxic suction cups. It is important to mount the heater clear of the gravel for two reasons: firstly, there must be adequate water circulation around the heater to prevent local boiling action occurring, which would crack the heater's tube; and secondly, many marine fishes lie on the gravel at night and could well suffer burns from a heater close to

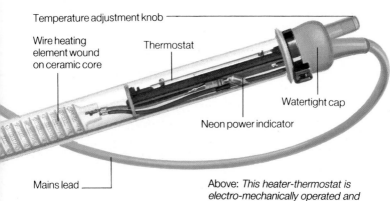

Temperature adjustment knob

Wire heating element wound on ceramic core

Thermostat

Watertight cap

Neon power indicator

Mains lead

Above: This heater-thermostat is electro-mechanically operated and magnetic algae scrapers should not be parked near it.

Above: *A single fluorescent tube of the standard domestic colour will provide sufficient illumination to view a 'clinical' type of aquarium.*

Above: *Several tubes, including independently switched 'actinic' types, will be needed to maintain lush algae and invertebrate life.*

their chosen resting place.

Always remember to turn off the mains supply before connecting or servicing any electrical equipment.

Lighting

When evaluating the light requirement of the marine aquarium, you have to take into consideration what type of aquarium you want (see the options discussed on page 14).

If you intend keeping a very clinical aquarium, where algae is not to be tolerated, you need only enough light to allow you to see the fishes, as any excess will undoubtedly bring the dreaded greenstuff! On the other hand, in the semi-natural aquarium you will want to approximate to the lighting levels found on the coral reef, and to provide lush growths of green algae on which the fishes will graze. (You may also need a brighter light if you use an algal-filter in your water-purification plans.)

For 'viewing only' lighting, a single fluorescent tube of the length nearest to that of the tank will suffice. For the semi-natural system, lighting levels of three or four times greater will be needed for efficient operation.

Fluorescent lighting is preferable to tungsten, for with the amounts of lighting required there will not be such

Below: *The latest 'hi-tech' lighting uses mercury discharge lamps to reproduce the power and brilliance of the sun's light over a tropical reef.*

a build-up of heat within the aquarium hood as is the case with tungsten lighting. It is vitally important that the hood is well-ventilated to disperse the heat from the lamps, whichever system is used.

Fluorescent lamps come in standard sizes (usually 10 watts per 300mm/12in length for normal diameter tubes) and in different 'colours'. For a single tube-lit aquarium you are stuck with whatever you start with, but in a multi-lamp system you can incorporate a mixture of tubes; a combination of 'warm-white' coupled with 'Grolux' or 'Northlight' gives an acceptable colour balance. Triton tubes have been specially developed for aquarium use and give a bright light that enhances the colour of the fishes and aids growth of marine algae. Soft corals, leafy algae and invertebrates all require high levels of lighting; under the influence of 'actinic' tubes many soft corals appear to fluoresce. It is usual to arrange independent control of each type of tube.

In special situations such as exhibitions, a marine aquarium can be lit to advantage by using high-intensity spotlamps. These give a dramatic and directional light, creating patches of light and shade,

and this shows up the water surface movement more than the overall even spread of light from fluorescent tubes. Of course, you can use a combination of lamps to give whatever effect you require, but beware of overheating the surface layers of the aquarium water.

It is normal for the aquarium to be lit for perhaps 10-15 hours each day, but the lighting level can be reduced from its full intensity to a lower level for evening viewing. Exact light requirements are best found by trial and error; in a semi-natural aquarium the aim is to have a reasonable growth of algae without it taking over the whole aquarium.

You must safeguard the lamps and their fittings from damage by condensation. Waterproofed connectors are available, but the best method of protection is given by using a cover glass placed on the top of the aquarium between the lamps and the water surface.

Below: *There is little chance of water spray or condensation damaging these lights; they have been mounted well above the water level of this slim-looking, all-glass marine aquarium. For lamps mounted close to the water level a cover glass is essential.*

Aeration and filtration

Water movement maintains the high oxygen content of the water – much desired by the fishes, which are used to such conditions on the coral reef – and also helps to drive out carbon dioxide from the water. Although water returning from filtration systems will agitate the water surface (as well as circulating the water around the aquarium), a powerful air pump will supply a constant stream of air to the diffusers or airstones (one at each end of the aquarium), and keep the water surface sufficiently turbulent. Airstones need not be the usual porous 'stones' (which are liable to clog up in sea water), but can be made from beechwood or sections of cane; replacements are easily made by the enterprising D-I-Y fishkeeper.

The air supply from the pump is also used to operate filtration systems and can be treated with ozone to help kill bacteria in the aquarium when used in conjunction with a protein skimmer (see page 28).

Filtration systems

It is critically important that the marine aquarium water is kept in tip-top condition and that the harmful effects of the fishes' waste products (visible and invisible) are reduced to a minimum. It is quite possible in freshwater fishkeeping to maintain a healthy aquarium solely by careful stocking and feeding, coupled with regular partial water changes, without any recourse to filtration systems. In the marine aquarium, however, water conditions are so critical that filtration *must* be used.

Almost any type of filter can be used in the freshwater aquarium, but the choice of a filter for the marine aquarium must be made carefully. Highly efficient filters, which leave the water crystal clear, may not be a good idea; this is not to say that the water should be allowed to become cloudy or discoloured, but such filters will remove many of the small organisms that filter-feeding fishes and invertebrates (if kept) rely on. Efficient filtration is still required, but you must keep the fishes' interests at heart at all times in making your final choice.

There are several types of conventional aquarium filtration

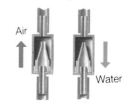

Top and above: *Fitting a 'one-way' check valve allows air through but prevents water returning to damage the air pump should it fail.*

methods that can be used, not necessarily singly to the exclusion of the others, and these include mechanical, chemical and biological processes. All these types can be operated either by air or by electrically driven impellers.

In addition, there are types of filters that seem to be used more exclusively in the marine aquarium and these include algal filters, ozonizers, protein skimmers and ultraviolet sterilizers. Details of the fitting and practical operation of the

filters will be given in a later chapter where necessary.

Despite the efficacy of modern air-operated outside filters, they are more suitable for freshwater use, because the majority of designs do not process sufficient water quickly enough for the requirements of marine fishes. The necessary rapid water-turnover rate is better supplied by power filters where the water is moved by impellers. Undergravel filters also require a fast turnover of water, but here the water flow *can* be handled by air pumps provided they are powerful enough and that the airlift tube is of sufficient diameter.

Mechanical and chemical filtration

As outlined above, the best type of filter is the external, sealed canister design. It usually has the motor and centrifugal pump mounted on the top, and will deliver a satisfactorily high flow of water. It can be hung on the side of the aquarium by means of a specially designed hanger, but it is more usually situated near the aquarium on a convenient shelf; because the system is sealed, water can be pumped to any level and for some distance.

Filter medium is placed in the canister body and removes the suspended matter from the aquarium water as it passes through the filter.

Below: *Water flow through undergravel filters can be increased by fitting a 'power-head' (below left) to each tube; some have a reversible action to permit reverse-flow usage.*

The most commonly used filter medium is some form of floss made from spun nylon, acrylic or some other man-made fibre. In order to prevent rapid clogging of the fibres it is normal practice to have some 'pre-filter' medium ahead of the floss in the water flow. This can be small pieces of ceramic pipe, but many fishkeepers use squashed together plastic pot-scourers.

Activated carbon is used to remove dissolved wastes from the water by adsorption. When it is placed in the filter body, sandwich it between two layers of floss to prevent it being drawn into the aquarium. Alternatively, it can be packed into a nylon bag, which will then serve the same purpose as the floss sandwich. Recently-developed resins can also be used to remove substances such as nitrate from the water.

The external filter can be used as a pre-filter to a reverse-flow undergravel filter, the principle of which is described below.

Whatever type of mechanical and chemical filter is used, you must renew the filter medium regularly; a permanently dirty filter medium will re-dissolve much of its contents back into the aquarium water.

Biological filtration

In this form of filtration, nothing is trapped nor adsorbed, and the cleansing process is done by living bacteria. The main toxic substances in any aquarium are ammonia and nitrite. Ammonia is excreted from the fishes' gills and is also the result of the breaking-down of their solid wastes.

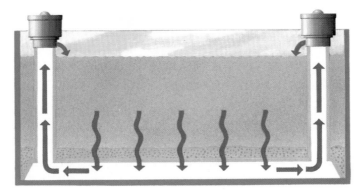

Nitrite is the halfway stage between ammonia and less harmful nitrate.

In the fishes' natural habitat the amounts of these substances are very low, due mainly to the huge dilution process provided by the ocean, and also because what little there is can be quickly utilized by planktonic algae and seaweeds almost as soon as it is formed. In the close confines of the aquarium, with no opportunity for dilution, the amounts are considerably higher and must be controlled.

The bacteria that perform this vital task live in the substrate of the aquarium. It is important that this area of the aquarium is kept well oxygenated, as the bacteria are *aerobic* and will die if exposed to *anaerobic* (oxygen-deficient) conditions. In undergravel filters a constant waterflow through the gravel ensures the correct conditions and this must be maintained at all times. Water can be either drawn down or pumped upward through the gravel, depending on the design of the system. To ensure a high water-flow rate, 'powerhead' electric impellers are recommended, as airlifts, even when fitted with wide bore tubes, may not provide enough flow.

In 'reverse-flow' systems an external power filter is used to drive pre-cleaned water underneath the filter and up through the substrate. While this method has the advantage of preventing early clogging of the substrate, it does not provide the beneficial water-surface turbulence of the alternative system.

The main attractions of undergravel filtration are that it is efficient, unobtrusive and needs very little maintenance. The undergravel filter is considered to be obligatory in the marine aquarium.

A biological filter system, such as a sand filter, can be set up outside the tank, forming part of a complex filtration plant incorporating other water treatment devices. Special filter mediums, allowing both aerobic and anaerobic bacterial filtering action, permit biological filtration within external canister filters. Semi-dry trickle filters (as well as sub-gravel anaerobic boxes) help to remove nitrogenous wastes completely by converting residual nitrates back in to free atmospheric nitrogen.

Other types of water purification

In addition to the methods of cleaning the aquarium water by means of the above filtration equipment familiar to freshwater fishkeepers, there are more advanced types of water

Below: *A power filter can be used to drive pre-cleaned water through the gravel bed of a reverse-flow biological filter. This will keep the gravel in the aquarium cleaner for a longer time.*

Below right: *Power filters can be positioned alongside the tank or on a nearby shelf. The gravel shown in this photograph is more suitable for use in a freshwater aquarium.*

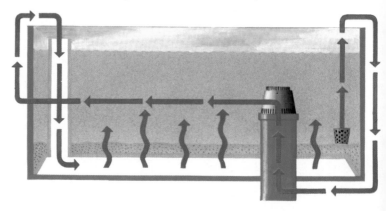

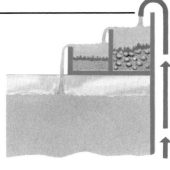

Above: *An external biological filter fitted above the aquarium. Water is fed to the gravel-filled tray through an airlift tube and spills back into the tank after flowing through the bacteria-rich filter. Useful aeration also occurs.*

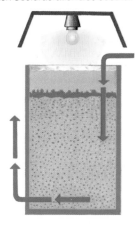

Above: *A layer of algae can be encouraged to grow on the surface of this external sand filter by the bright light above it. Such algae filters can reduce carbon dioxide.*

treatment devices. Some of them will provide very clinically clean conditions but may not be entirely suitable for those wishing to maintain a more natural aquarium environment. They could be used to advantage in quarantine or disease-treatment aquariums.

Algal filter

Aquatic plants utilize carbon dioxide in their photosynthesis processes and do much to keep this unwanted gas down to a minimum. The marine aquarium relies much upon water movement to expel carbon dioxide from the aquarium, particularly at the water surface. Unfortunately the marine aquarium is not blessed with a profusion of water plants, as is the freshwater aquarium, but it is possible to reduce the carbon dioxide levels by the use of the most primitive of plants – green algae. Although in theory this is quite reasonable, the practical applications are not without problems, for the amount of algae required to be of effective use is quite considerable, and the algal filter must consequently be a large affair.

An algal filter can take two forms: an outside sand filter can be encouraged to produce an algal top layer by shining a bright light down on its surface; or the water from another type of filter can be returned across a shallow well-lit tray on which algae are again encouraged to grow. Algal filters are not commonly found in aquariums of home proportions, but may be incorporated in much larger aquariums where there is sufficient space to accommodate the trays above the aquarium.

Protein skimmers

If air is bubbled through water containing a high amount of organic matter, a scummy foam will form on the water surface; this foam will contain much of the organic matter, which can be collected and removed.

The protein skimmer is a device that works on this principle. Aquarium water passes through a vertical tube, during which time it is subjected to a column of air bubbles; the air bubbles create foam, which is then forced up through an exit at the top of the

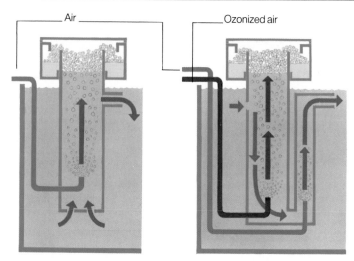

Above: *In a direct-current protein skimmer tank water enters at the bottom and flows in the same direction as the stream of air bubbles. Water-flow through the skimmer is fairly fast and exposure to the air or ozonized air is relatively short.*

Above: *In counter-current designs of protein skimmers, the water-flow goes against the current of air or ozonized air and is therefore much slower, thus giving the water more exposure to the effect of the air bubbles or ozone sterilization.*

protein skimmer into a collecting chamber, where it settles out into a yellow liquid. It is then a simple matter to empty the liquid from the chamber.

There are two types of protein skimmer. In one, the water enters the skimmer at the base and exits at the top through an overflow outlet; in fact, it may be regarded as nothing more than an enlarged modified airlift tube, fitted with a collecting chamber on the top. The water is subjected to the foaming action for only the short time it takes the bubbles to rise the height of the

tube. This type of skimmer is known as a *direct-current* device, the water following the same path direction as the air bubbles. There is usually only one airlift, providing both water movement and foaming action, in skimmers of this design.

An improvement on the design is found in *counter-current* models. Here, water enters the tube at the top and leaves at the bottom through a separate airlift tube. Another airstone provides the foaming action, and as the water moves down, against the current of the foaming bubbles

Below: *Check valves can be used as shown to protect air pumps should they or the electricity fail, but ozonizers should be protected by an 'anti-siphon loop' in the airline instead because the ozone will seriously damage the check valve.*

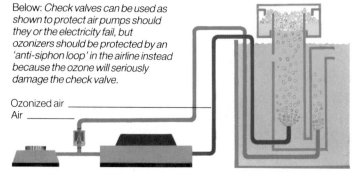

Ozonized air
Air

coming up, it moves more slowly and so is subjected to the foaming action for a correspondingly longer period of time. Methods of collection and disposal of the foam are unaltered.

A 'powered' version of protein skimming works slightly differently. Its diffusion principle allows collection and disposal of protein but without removing any planktonic food (vital to filter-feeders).

Using an ozonizer

Ozone – an unstable form of oxygen containing three atoms (rather than the normal two) – is a very powerful bactericide. It is created by passing air from the air pump through a field of electrical discharges in a device known as the ozonizer.

If you are using an oil-lubricated piston pump, the air should be filtered to remove traces of oil before it enters the ozonizer; air from vibrator pumps can be dried by using a silica gel filter before the ozonizer to remove any water vapour. The output from the ozonizer, containing air and ozone, is fed to the protein skimmer.

There are attendant dangers in using ozone. Always use plastic airline tubing, as ozone will severely damage rubber tubing, pump diaphragms and check valves. Ozone is also very poisonous and should never be fed directly into the aquarium water, where fishes may come into contact with the bubbles.

This is another good reason why it is best to use ozone in conjunction with the protein skimmer. If you incorporate ozonized air into any reverse-flow undergravel airlift, it will destroy the nitrifying bacteria. It is best to exclude the use of ozone from aquariums containing filter-feeding invertebrates.

Ultraviolet sterilization

Like ozone, ultraviolet light is a powerful bactericide and must be used with caution.

The ultraviolet light is generated by a special lamp whose 'glass' envelope is actually made from quartz (do not touch it with the naked hand). The lamp is built into an outer, darkened jacket through which the aquarium water is passed. Do not attempt to test the ultraviolet lamp outside its jacket, and never look at it directly without suitable goggles.

The efficiency of the ultraviolet sterilizer is dependent on the rate of water passing through it and the intensity of the lamp. Again, aim for the maximum time of exposure of the water to the light. As the sterilizer is mounted externally, there is no danger of the fishes being harmed, but any water-borne bacteria and micro-organisms (good and bad) will be killed in the process.

One result of keeping fishes for prolonged periods in clinically sterile conditions may be to condemn them to those conditions for ever, as they may eventually lose their natural resistance to disease and become unable to withstand the shock of being transferred to the 'bacteria-ridden' conditions of the more normal, natural-looking aquarium.

Below: *A powerful ultraviolet lamp housed in a special quartz jacket irradiates water that passes around it, sterilizing it completely.*

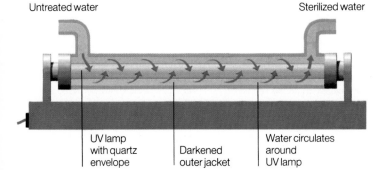

Untreated water   Sterilized water

UV lamp with quartz envelope | Darkened outer jacket | Water circulates around UV lamp

Water

Typical salinity strengths

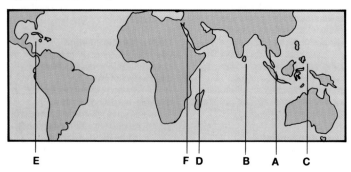

A	Singapore (Indian Ocean)	**30-32 gm/litre**	*The salinity of*
B	Sri Lanka (Indian Ocean)	**30-34 gm/litre**	*sea water varies*
C	Philippines (Pacific Ocean)	**30-34 gm/litre**	*around the world, the*
D	East African Coast (Indian Ocean)	**32-35 gm/litre**	*most land-locked*
E	Caribbean Sea (Atlantic Ocean)	**35 gm/litre**	*seas registering the*
F	Red Sea	**40 gm/litre**	*highest salt content.*

Success in marine fishkeeping is directly proportionate to the ability of the fishkeeper to maintain the aquarium water at its optimum condition. Apart from the temperature required for keeping tropical species, there is also the salinity of the water to be considered.

Salinity

The salinity of natural sea water varies very little throughout the oceans of the world. The only factor affecting the strength of the sea water is in localities where the formation of the land prevents regular water movement. Hence, the Mediterranean is saltier than the Atlantic Ocean. The Red Sea is the source of many of our aquarium species; open to the Indian Ocean at one end only, and subject to rapid evaporation, it has a salinity higher than that of open ocean water.

Landlocked sea water that has no freshwater discharges into it, and also registers a high rate of evaporation, will record a very much higher salinity still; a classic example of this is the Dead Sea, in which neither animal nor plant life are viable now.

You will see from the table that it would be wise to enquire of your aquatic dealer about the source of the fishes that are offered for sale, so that they can be acclimatized to the salinity conditions in your aquarium.

Testing salinity

The salinity of aquarium water is tested by means of a hydrometer, a simple device that indicates the density of the water in which it floats by the amount to which it sinks or rises. The lower it sinks, the less dense the liquid (lower salinity); conversely, a high floating level indicates a higher salinity.

The hydrometer, which must be calibrated for the ranges found in the marine aquarium, has a scale of specific gravity shown on its stem; where the water level crosses the scale is the indicated specific gravity (S.G.) of the water. Specific gravity is the ratio of the density of any liquid to the density of distilled water. Measurements of specific gravity should be taken at the same water temperature, as a rise in temperature will result in a lower S.G.

Typical S.G. readings for a marine aquarium range between 1.020 and 1.025, but bear in mind that at the higher S.G. there will be a greater strain put upon the metabolism of the fish, as it constantly fights against losing water from its body (page 13).

Sources of sea water

There are two sources of water suitable for the marine aquarium: the sea itself, and the synthetic sea water that you mix yourself, using salt mixes.

If you live close to the sea and intend to keep fish species from that area, you may find that – provided you have access to a *clean* source of sea water – you can successfully keep fishes in natural sea water. The problem is, of course, that natural sea water (despite being theoretically the most suitable) is too full of potentially dangerous organisms to be considered totally reliable for aquarium use, even if the collection and transport of sufficient quantities present no difficulties.

Synthetic sea water has many practical advantages: it is convenient, available from your local store, and completely hazard-free. Follow the instructions exactly when mixing the salts; it should be unnecessary to stress that metal containers should not be used!

Maintaining water quality

As soon as living organisms are present in water, its quality begins to deteriorate. Efficient filtration and good aquarium management do much to slow down this process but cannot halt it entirely. Nitrifying bacteria in the biological filter will oxidize ammonia, first to nitrite and then further to nitrate. Water can be periodically tested with a nitrite test kit, to see how effectively the bacteria are coping with their important task.

The build-up of nitrate can be minimized by regular (monthly) partial water changes of 25-33%, or by using a denitrification system.

The replacement water must be of similar S.G. and temperature to that which it is replacing. One of the new filtration media may also be used to reduce nitrate in the water.

Water losses caused by evaporation should be replaced with *fresh* water, as salts are not lost during evaporation.

Keeping a check on pH

The pH of the water (its acidity or alkalinity) should be between 8.0 and 8.3, and can be conveniently determined with a pH test kit. (Make sure that you use a saltwater pH test kit; freshwater types, although covering the relevant range, will not show a true reading.) A falling pH – a normal tendency in the marine aquarium – can be counteracted by using calcium-rich coral sand as a substrate. Alternatively, pieces of seashell (another source of calcium) can be placed in the external filter.

The lowering of the pH is due to a build-up of carbon dioxide, decomposition or organic matter, and an increase of acids and nitrates. Often a falling pH means only that the water is getting old and is due for a change; but however bad the water is, do not try to make amends by changing a large proportion of the water at once. A gradual replacement is ideal, to avoid stressing the fishes.

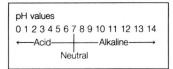

pH values
0 1 2 3 4 5 6 7 8 9 10 11 12 13 14
←———Acid————|————Alkaline———→
            Neutral

Below: *Two ways of testing specific gravity; with a floating hydrometer (left); a 'swing-needle' unit (right).*

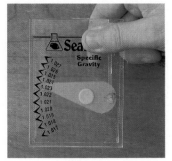

Aquarium decoration

At first glance, decorating the marine aquarium may seem an impossible task, for where are the lush aquatic plants? Fortunately, the undersea world has its own share of decorative materials, some living and some dead. In our semi-natural aquarium, we shall concentrate on purely decorative and functional decorations, rather than include coverage on the extra considerations posed by keeping living corals, tube-worms and other invertebrates.

Choosing the substrate
The covering of the aquarium base also doubles as a biological filter bed and water conditioner.

The provisions that prevent the selection of some gravels for use in the freshwater aquarium do not hold here. Any substrate containing calcium carbonate (a well-known water-hardening agent) will be generally beneficial in maintaining the water's correct pH. In aesthetic terms, ordinary aquarium gravel or fairly coarse river sand hardly fit the bill, as their colours may be completely out of keeping, but they can be disguised by a layer of natural crushed coral or coral sand on top of the gravel. This softer top layer will be appreciated by those species that burrow into the substrate at night.

Whatever the material used – coral sand or oolite are ideal – the substrate should be sufficiently deep to allow efficient biological filtration to occur. A depth of at least 5-7.5cm (2-3in) is recommended, and the substrate can be sloped from the front up to the rear of the aquarium to an even greater depth. This not only makes it look interesting, but also causes dirt and detritus to collect at the front of the aquarium where it is quickly noticed and removed.

Crushed coral or coral sand can also be used in the external filter system as a pH stabilizer and biological filter.

Using corals in the aquarium
A feature of the marine aquarium is the use of coral heads, both for decoration and as a refuge for the fishes. Only one coral will retain its natural colour when dead – *Tubifora*, the red Organ Pipe Coral; the skeletons of other species turn white as the colour-giving organisms die.

Do not overcrowd the aquarium with a tangle of coral, but try to make attractive arrangements in the tank.

Always disinfect, clean and bleach coral (if necessary) before using it in the aquarium. Much imported coral is only superficially cleaned and may have dead animals still clinging to it.

Seashells

Much of what has been said about coral applies to seashells and their use in the aquarium.

Rocks

Any rock containing obvious metal ore-bearing veins should not be used in the aquarium. Slate, limestone and hard sandstones are quite safe, and well-weathered rocks from the seashore are especially suitable.

Seaweeds

The culture of seaweeds in the aquarium is very much in its infancy. The successes that have been achieved appear to be limited to species of the marine alga *Caulerpa*. Other species, such as *Ulva*, *Enteromorpha* and *Cladophora*, might be tried, but may need even more light or depend on extra carbon dioxide being supplied for their successful culture in the aquarium.

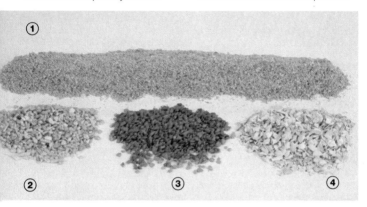

Above: *Substrate materials for marine aquariums. 1 Coral sand for the topmost layer. 2 & 3 Optional nitrite removers. 4 Crushed seashells to stabilize the water's pH.*

Below: *This well-established aquarium needs plenty of light to maintain its abundant green growth. Using the correct substrate materials will help to stabilize water conditions.*

Setting up

The tropical aquarium

A practical sequence of setting up the tropical marine aquarium is as follows:

1 Coral preparation
2 Water preparation
3 Site and tank preparation
4 Fitting biological filtration system
5 Substrate and decorations
6 Heating, aeration and other filtration equipment
7 Lighting
8 Adding the water
9 Start up and final checks
10 Tank maturation

It may seem that this order is contrary to expectations, but there are good practical reasons, as you will learn.

It is always a good idea to make sure that everything you are likely to need (even if some items will be used only once) is within easy reach. A table nearby for tools and equipment will be a blessing, as will some advance plan of action. With the size of aquarium necessary for marine fishes, it is unlikely that any advance preparatory work on the tank itself will be possible, so the aquarium must be set up from scratch in its final location.

1 Coral preparation

Decorative coral is made from the skeletal remains of the millions of polyps that make up parts of the coral reef. As it is not living, you must ensure that there are no remains of the dead animals in it before you put it into the aquarium. It takes a relatively long time to ensure that coral is sufficiently clean for aquarium use. There are two ways to clean it.

One is to soak it in household bleach (2 cups per 5 litres/8.8pt of water) for one or two weeks. Following this, soak it in several changes of fresh water for another similar period, until all the smell of bleach has disappeared. Where the coral has a highly convoluted surface it should have several repeat cleanings. This method is fairly popular, but sometimes it is difficult to get rid of the smell of bleach, once the coral is clean!

Alternatively, the coral can be boiled for an hour or so and then

thoroughly rinsed down with a hose to remove any loose particles. An overnight soak should be followed by another hosing-down and then the coral can be naturally bleached in the sun for a few days before use.

2 Water preparation

Synthetic sea water must be strongly aerated for at least 24 hours before being used in the aquarium; it should preferably also be stored in a dark container during this time.

Before mixing the salt, you can add dechlorinating agents to the bucketful of tapwater to remove chlorine, chloramines and heavy metals.

A practical problem immediately arises – in what do you mix the sea water? Obviously if it is to be your first aquarium the tank itself can be used, but here again there can be complications. If you use the tank and make a tankful of water, where does the surplus water go to when you put in the substrate, decorations and equipment – apart from over the floor? Mixing the water in the tank after the decorations etc have been put in may be easier, but the carefully arranged substrate and decorations may well be disturbed.

Calculating the net amount of water required for a furnished aquarium is difficult; it is probably

Below: *Take care to use non-toxic, non-metallic containers and stirrers when mixing synthetic salt water. Mix up all the salts in the packet.*

Above: *The 'life-support' equipment is cleverly hidden in cupboards beneath this stylish marine tropical aquarium. The aquarium frame is made of non-rusting stainless steel.*

better to make up water of the correct S.G., to the total volume of the aquarium, in a separate container. Be prepared to waste some water.

When mixing the water, there are two things to remember: use plastic containers; and always use *all* the salt mix. The latter point ensures that all the necessary trace elements are present in the resulting sea water; if you use only half a bag of salt mix there is a chance that some of these trace elements will be left in the bag. Another reason for using all the salt at each mixing is that any salt left, perhaps in an ill-closed bag, will absorb moisture from the atmosphere and this can adversely affect any salt-weight calculations in future mixing sessions. In any case, follow the instructions exactly.

Once the water has been mixed, a plastic dustbin makes a good container; it should be stored away from the light and aerated strongly for at least 24 hours. This will ensure that the salt is completely dissolved and that the chlorine from the tap water is

driven out of solution. Synthetic water can be stored in sealed plastic jerry-cans until needed.

There is no point in testing the specific gravity of the water at this time, as it will not necessarily be at the correct temperature; final adjustments can be made when it is added to the tank, by adding salt to increase the S.G. or *fresh* water to lessen it.

3 Site and tank preparation

The tank site area should be clearly accessible during the setting-up stage, and there should be no obstacles to trip over – especially when carrying buckets of water or valuable corals. Make sure that you have enough airline and electric cable, connecting blocks, air valves and electrical plugs before you start.

The aquarium stand or site must be firm and level, with nearby access to an electrical power outlet. Fit a Residual Current Device for extra safety. There should also be enough room around the tank to be able to maintain it. A shelf above or below will be used for some of the external equipment (they can always be boxed in afterwards). Remember to place expanded polystyrene on the site surface first, so that the tank sits on a cushioned surface.

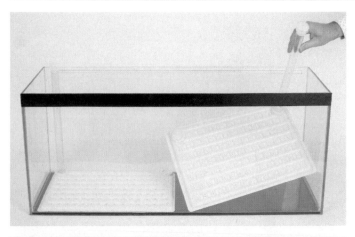

New tanks should not leak, but if in doubt test yours (outdoors!) beforehand. Leaks can be simply repaired with aquarium sealant, but again this is an outdoor job, as the sealant gives off a heavy vapour.

A cable tidy and a bank of ganged air valves can be stuck to the outside of the tank in convenient places, but we shall not be making use of them just yet in setting up.

Before you place the tank in its final position, you can paint the outside rear wall so that the room wall decorations will not show through the tank and spoil the artistic underwater picture. Alternatively, if there is room behind the tank a dry 'aquascape' of corals and sea fans can be built up, which will apparently extend the visible rearward 'depth' of the tank.

4 Biological filtration system

The biological filtration system must be fitted into the aquarium before the substrate is added. Make sure that the filter plate covers the whole of the aquarium base. This ensures the maximum area of filtration, and also that there are no anaerobic pockets. (To make absolutely sure that there are no 'short circuits' around the edges of the filter plate for the water flow to take rather than be drawn through the substrate, you can seal the filter plate into position with silicone sealant; but you must then allow 24 hours for it to cure.) Fit the airlift tubes and make sure that they are firmly attached. If the airline is connected right at the base of the airlift, now is the time to fit it, as later it may well be under the surface of the

Far left: Two subgravel filter plates may be needed to cover the aquarium base. Put these in at the beginning of the setting-up process.

Left: Ensure that there is an adequate depth – at least 5cm (2in) – of substrate (in this case coral sand) covering the entire filter plate area.

Below left: Use pieces of coral or inert rocks to hide the 'hardware' in the tank. Note that a check valve has been fitted in the airline.

Below: Handle corals with care: they are very brittle. Make sure that corals are thoroughly clean before use.

substrate and much harder to get at. Connect the other end to an outlet on the ganged air valve block.

In the case of power-assisted biological filters, the motor impeller head can be fitted to the top of the return tube later; there will be a slight problem in using these with a normal reflector/hood, which may need some physical modification to accommodate the power head, as will the cover glass.

5 Substrate and decorations

All materials used as substrate should be well washed before use. This is particularly important if crushed coral or coral sand is used, as these may contain dead organisms. It is probably not worth the effort to test it (by soaking under water for a few

days – your nose will then tell you if it is clean!), and quicker to wash it automatically in the course of events. At the same time also clean the rocks.

Place any large rocks and corals directly on the filter plate, and then spread the substrate medium over the biological filter plate to an even depth of about 5-7.5cm (2-3in). Add more substrate to increase the depth towards the rear of the tank by a further 5cm (2in) or so, using small rocks under the substrate to hold back the slope.

Finally, add corals and further rocks as required (they can usefully hide the aquarium 'hardware'), but do not take up too much of the fishes' swimming space nor leave any areas of the aquarium obscured where a fish could become trapped.

6 Heating, aeration and other filtration equipment

The heating equipment is best mounted at each end of the aquarium, whether you are using combined heater/thermostat units or separate heaters controlled by one or more external thermostats. Make sure combined units are mounted almost vertically and check whether they are designed for complete immersion in water. Mount heaters clear of the gravel, and always use plastic mounting clips for all equipment. Connect the wires from the heaters/thermostats as guided by the manufacturer's instructions, and wire both these and the air-pump supply cables to the appropriate connections in the cable tidy. DO NOT CONNECT THE MAINS SUPPLY AT THIS TIME.

The output from the air pump should be fed into the air valve block. The pump itself should ideally be mounted above the level of the water in the aquarium to prevent any possibility of water siphoning back into it, should the electricity supply fail or the pump stop for any other reason. Alternatively, a 'one-way' check valve can be fitted in the airline from the pump, which serves the same purpose and allows the pump to be sited where most fishkeepers put it anyway – on a convenient shelf near the aquarium.

A long airstone can be placed along the back of the tank and hidden from view by a rock or coral heads. The airline tubing from the airstone should be connected to one of the air valves, and can be effectively hidden in the aquarium by burying it in the gravel or hiding it behind rocks.

Fill the external filter with filter medium as required. External filters, and ultraviolet sterilizers can be situated remotely from the tank or hung on the tank side or rear. Cut the hood away to allow the necessary tubes to pass into the tank. The returning water should be distributed across the surface of the aquarium by

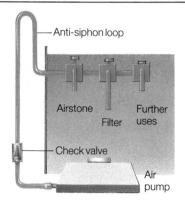

Above: *Distribute pumped air to the aquarium appliances via ganged air valves. Note the 'back-siphoning' precautions taken for the air pump.*

means of a spray bar; apart from avoiding a powerful jet of water disturbing the tank decorations, it will help to disperse carbon dioxide and also aerate the water, as will the output from power filters fitted with regulated air controls.

Protein skimmers are best fitted unobtrusively as possible in the corner of the aquarium. Their supply of ozonized air should come *directly* from the ozonizer, which is fed with air from an outlet valve on the ganged air valve. The ozonizer should be safeguarded against back-siphoning by using an anti-siphon loop, as check valves can be damaged by ozone. Remember to use only plastic tubing for the passage of ozonized air.

Right: *Combined heater/thermostats should be mounted vertically, but if the water depth should prevent this then position them as shown here.*

Connect the power supply cables from external power filters to the cable tidy. The sterilizer and ozonizer will not be needed immediately and their connection can be left until later on in the setting up procedure if desired.

7 Lighting
The lamps should be fitted into the hood with waterproofed fittings. The associated starting gear for fluorescent lamps is very heavy and can make the hood unbalanced if fitted therein; be careful not to drop the hood on the cover glass. Where a number of tubes are used, the starting gear should be mounted remotely from the hood.

Below: *The fluorescent tube shown here (although fixed too far back in the hood) has ideal waterproof connectors. Mount the starter gear for the lights near the aquarium.*

Ensure that the hood is well ventilated. A cover glass will also keep evaporation losses to a minimum and prevent fishes from jumping out of the tank.

Connect a power lead to the switched terminals in the cable tidy.

8 Adding the water
Once all the tank decorations and equipment are in place the water can be added. To avoid disturbing the contoured gravel, pour the water onto a saucer resting on the substrate; the overflow will gently fill the tank without causing any disturbance. When the tank is almost full it is time to connect up the electricity supply and switch on. The small amount of unfilled tank space is to allow for adjustments to the water's specific gravity. This is best carried out when the water has reached the correct operating temperature.

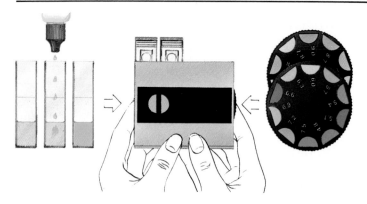

Above: *This pH and nitrite test kit uses liquid reagents to cause colour changes in separate water samples; comparison to reference wheels viewed through a 'control' sample of water gives direct pH and nitrite readings. Be sure to use only those wheels calibrated for sea water.*

9 Start up and final checks

It will be necessary to prime the outside filter and UV sterilizer circuits with water. This simple process is best achieved by removing the return tube to the aquarium at the appliance end and sucking at the outlet from the filter or sterilizer. When water has filled each system, reconnect the return tubes, making sure that all water hose connections are tight. An airlock in the system can generally be removed by gently rocking the filter in question to manoeuvre the trapped air round to the outlet.

Connect the power lead from the cable tidy to a suitably fused plug, and plug into the power socket and switch on. Adjust the air flow from the air valves to operate the biological filter and airstone (if fitted). Check the water-flow from any external filter.

Check the aquarium thermometer to confirm that the heating apparatus is working. When the temperature is up to the set level (approximately 24°C/75°F) test the specific gravity of the water with the hydrometer. If it is low, add some more salt (dissolve some in water before adding to the tank) and let the water circulate for a while before rechecking. Continue until correct. If the S.G. is high, add some *fresh* water as needed.

10 Maturing the filter bed

In the initial period after setting up, the bed of nitrifying bacteria will take some time to become established. There will be a build-up of ammonia and nitrite, which will gradually diminish as the bacteria get to work in ever-increasing numbers. The maturation of the filter bed takes from two weeks to more than a month, depending on what materials are available for the bacteria to work on.

Maturation can be hastened in several ways, all ensuring that there is a supply of nitrogenous compounds to act as fuel for the system. You can add some substrate from an established marine aquarium, allow small particles of food to rot away, or put one or two relatively hardy fishes in the aquarium to provide the basic materials themselves.

Testing the water with a nitrite test kit will show how the maturation is progressing; when it reaches a stable minimum level, it is safe to introduce some more livestock.

The coldwater aquarium

It will be assumed that although you are serious about setting up a coldwater marine aquarium, the fact that all the 'ingredients' are fairly easily obtainable (and replaceable) will mean that you will not be too upset if it does not turn out as successful as you originally hoped.

The main thing is to collect sensibly, taking special care not to damage any of the animals' natural homes in the process. A little has already been said about the dangers of collecting polluted sea water.

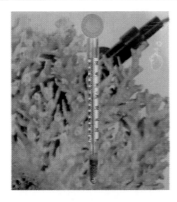

Above: *Fix thermometers where they can be easily seen, ideally with plastic suction caps. Spirit or mercury types are the most accurate and reliable.*

Collect water away from beaches, during a settled period of weather (recent heavy rain or storms may result in relatively diluted and dirty sea water). When collecting from a boat, take water from the side of the boat that is meeting any tide or currents; in this way you will not collect any oil from the boat. Use plastic containers to carry the water home. Filter the water to remove any dying plankton and clean the filter frequently during the first few weeks.

The fishes and animals collected

Below: *Some coldwater marine invertebrates, such as crabs, need 'above water' areas they can climb on to. Build up a corner with rocks.*

from the rockpools must not be crowded while you carry them home. If you collect invertebrates such as sea anemones, always gather them together with the rock on which they are sitting; prising them off may damage them. Many of the inhabitants of the pool rely on rocks for living quarters, so always replace rocks removed with substitutes.

The collection of seaweeds is questionable. Marine algae need a great deal of light, and if you provide this there is a danger that you might overheat the aquarium.

Setting up a coldwater aquarium must be a reasonably swift operation if it is to be done at the end of the collecting trip, unless you make arrangements to collect the substrate and water during a previous visit to the coast, returning later for the livestock. As many of the fishes will require hideaways, you must also incorporate such shelters in the furnishing design of the aquarium.

Setting up follows the above general pattern, but there will be no need for any heating apparatus. Indeed, some cooling may be necessary during warm periods. It will be better for any filter-feeding animals if powerful external filtration is not used; a biological filter (combined with extra aeration) will suffice.

The design of the aquarium layout may be modified if necessary, to provide a portion of 'land' above the water level, on which crabs might exercise; fit a cover glass!

Feeding

Marine fishes have a wide range of methods of feeding: some will eat anything, others are strictly vegetarian; predatory fishes like to eat moving – preferably live – foods, whereas the grazing species search the surface of the tank decorations for tidbits. The aquarium may contain meat-eaters, alongside those species that eat very fine particles of food. On top of this, you then have the added problem of catering for the hearty eaters together with the more nervous and shy species.

Live foods

Just as in freshwater fishkeeping, the value of live foods cannot be underestimated; those of marine origin may well contain additional beneficial trace elements. An important reason for using aquatic crustaceans as food is that they provide valuable 'roughage' to aid digestion. Many live foods from freshwater sources (*Cyclops*, *Daphnia*) can be used as food, but you must realise that these will live for only a short time in the marine aquarium. Try to feed the correctly calculated amounts; this will save you having to net out the dead surplus

Below: *Some natural fish foods suitable for the marine aquarium.*
1 Mysis shrimps. 2 Lobster eggs.
3 Cockles 4 Lancefish. 5 Prawns (Cerastoderma sp.). These may be stored in the freezer until required, but do ensure that they are thoroughly thawed before use in the aquarium.

after the fish have finished feeding. Young freshwater fishes such as live-bearer fry will be taken by some fishes (Seahorses are said to be quite partial to newly born Guppies), but always remember that their lifespan in salt water will be short.

The larvae of mosquitoes, midges and gnats are available in the summer months from garden rain-butts. These will be eagerly accepted by many fishes.

Tubifex worms, if properly cleaned (keep them under running water before use), make an excellent fish food. They will burrow into the substrate, however, which may cause problems, because they soon die and will pollute the aquarium. If you intend to use *Tubifex*, feed sparingly, use a floating worm-feeder and remove any surplus worms immediately the fishes lose interest.

The same feeding precautions should be taken with other live worm-foods, such as white worms and grindal worms, which should not be given too often, because they are extremely fattening. Earthworms are excellent food but again need to be cleaned before use. If you have to shred them for smaller fishes always give the pieces a good rinse under running water, to get rid of the earth and blood, before feeding.

Saltwater live foods include the shrimps *Artemia salina* and *Mysis*. The former, known as the brine shrimp, is available in egg form, which can be hatched in warm salt water to provide a disease-free nutritious food

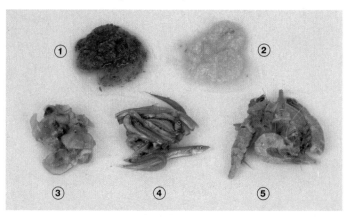

for fishes. They also make good food for filter-feeding invertebrates. It is possible to raise brine shrimps from the newly hatched nauplius stage to adult size when they are then usable as food for larger fishes. This process can take a seemingly long time (perhaps several weeks), with the shrimps requiring feeding with special yeast-based foods; it is more convenient to buy living *Mysis* shrimps, which are often obtainable from your aquatic dealer.

The lugworm, found in beach sand or mud, also makes a good source of food for marine fishes.

Meat foods

It will come as no surprise to learn that a favourite food of fishes is fish. Lean white fish meat and the meat from shellfish (crabs, mussels, prawns, shrimps) is very useful in varying the fishes' diet. This is important, because it stops them becoming bored with the same food, and also provides a means of supplying vitamins that might otherwise be

Above: *Many marine fishes are herbivorous. By encouraging lush growths of marine algae, such as* Cladophora, *in a well-lit aquarium you will ensure they are adequately fed.*

missing from regularly given foods. Crushed snails, either from a freshwater aquarium or from the garden, are also accepted.

Lean beef and ox heart scrapings also help to vary the list of foods, and liver will provide vital vitamins.

The home freezer can be used to good effect to store frozen seafoods, or home-made recipes, for future use. Be sure that the food is thoroughly thawed before feeding it to the fishes.

Vegetable foods

Many marine fishes are grazing animals, requiring vegetable matter in their diet. This can be supplied by making sure that there is a healthy growth of algae in the aquarium, or by supplementing the food given with lettuce, spinach or pre-soaked rolled oats. The lettuce and spinach can be

lightly crushed or even cooked a little
to soften them. Once fishes learn that
these green leaves are food, the
leaves can be floated in the aquarium
or anchored to the bottom with a
weight: a very easy way to do this is to
clamp a lettuce leaf between the two
pieces of a magnetic algae scraper,
but make sure the scraper is nowhere
near the aquarium thermostat; it may
disturb its operation.

Freshwater aquatic plants can also
be used as food; algae scrapings can
be transferred to the marine
aquarium, and water wisteria
(*Synnema triflorum* or *Hygrophila
difformis*) and *Riccia fluitans* are also
useful in this respect.

Manufactured foods

Many manufacturers of food for
freshwater aquarium species have
also turned their attention to the
provision of foods for marine species.
A wide variety of dried flake foods is
commonly available. There are also
frozen foods, primarily of sea foods
such as shrimps, that have been
irradiated with sterilizing gamma rays
to protect your fishes from any risk of
disease being introduced with their
food. In addition, many of the live
foods outlined above are available in
freeze-dried form, which brings them
conveniently to your aquatic dealer's
shelves. They appear not to lose any

of their nourishment in the freeze-
drying process. You will have to rely
on the water currents to give these
dead foods a 'lively' character when
placed in the aquarium!

Feeding techniques

Apart from the golden rule – NEVER
OVER-FEED – there are only one or
two guidelines to follow. Many marine
fishes are continual feeders, although
some of the more predatory species
will space their meals out, so it is a
case of 'a very little and very often'
rather than having set mealtimes.

This procedure should be adopted
once the marine aquarium has
become established and the fishes
are feeding normally. During the initial
settling-in period, the fishes will not
feed so willingly and any inadvertent
over-feeding will result in the
biological filter becoming overloaded.

Also during the first few weeks, it is
advisable to feed predominantly live
foods in order to get the fishes
feeding regularly as soon as possible;
otherwise, by the time they have
settled down, they could well be on
the way to starving to death if they are
not given sufficiently tempting foods.

The rule of feeding all the food that
the fishes will eat in a few minutes is
not always easily applied to the
marine aquarium. Because of the
various feeding habits of the fishes,

Above left: *Clamping a lettuce leaf between the two parts of a magnetic algae scraper is an easy and convenient way to anchor it down so that fishes, such as this Yellow Tang* (Zebrasoma flavescens), *can browse on it. Remove uneaten fragments.*

Above: *A starfish (*Protoreaster *sp.) feeds on a dead Guppy provided by its owner. Digestion occurs in the central stomach and in the arms.*

you must supervise the feeding until all the fishes have been seen to take food – there is a big difference between the fishes being fed and the fishes actually feeding. Once the bolder fishes have satisfied themselves, then persevere with a little more food for the more shy species. Always remove any uneaten food immediately.

Grazing fishes can be effectively catered for by painting a rock with a liquidized 'emulsion' of meat and algae foods. When dry, the rock can be put into the tank and the fishes will be attracted by the food on its surface just as they are in nature.

Types of feeders
The following is a list of the feeding habits of the majority of fishes kept in the marine aquarium. It should be used as a rough guide only, for a

species kept in a variety of collections may well exhibit different characteristics depending on the attitudes of tank-mates.

Bold feeders:
Blennies,
Clownfishes, Damselfishes,
Fingerfishes, Gobies,
Rabbitfishes, Scorpionfishes,
Snappers, Squirrelfishes,
Triggerfishes

Vegetarian grazers:
Angelfishes,
Butterflyfishes, Filefishes,
Parrotfishes, Surgeons,
Wrasses

Shy feeders:
Basslets,
Cardinalfishes, Croakers,
Hawkfishes, Jawfishes

Filter feeders:
Pipefishes,
Seahorses, Shrimpfishes,
Trumpetfishes.

Species can fall into more than one group: for instance, many bold feeders require algae or other vegetable matter in their diet. Further information on individual feeding needs will be given in the Species Catalogue later in this Guide.

Choosing the fishes

Marine fishes are much more delicate than their freshwater counterparts, and less able to cope with varying aquarium conditions. Because they are also wild-caught and not commercially bred in aquariums or fish farms, they have not developed any hardiness and are often quite debilitated by their long journey to the aquatic dealer's tanks. In addition, they may be suffering from the effects of cyanide-based drugs used in their capture in the wild.

Bearing all these things in mind, in order to stand any chance of keeping these fishes successfully you must do everything right – there is very little margin for error. You will give yourself and the fishes a head-start if you select only healthy stock.

Healthy or not?

There are two important considerations when choosing fishes: health, and aquarium compatibility. A third complication is that some of the most desirable fishes, carefully chosen using the first two criteria, very often disappoint by being difficult to feed. The dealer's advice should be sought in this case (see below).

Appearance and behaviour tell you a lot about the fish's probable state of health. Colours should be dense, with clearly marked patterns; dull patches indicate over-production of mucus, often a sign that something is wrong. A patch of missing mucus, on the other hand, caused perhaps by clumsy or incorrect handling during transit, leaves the fish open to infection in the aquarium.

Rapid breathing also shows that something is wrong – parasitic disease, lack of oxygen or bad aquarium conditions. The fish must be able to swim effortlessly, and any bobbing about should be taken as a danger sign. Freshwater fishkeepers expect to see healthy fishes swimming about with fully erect fins; but marine fishes often swim with

their fins folded. Some species have no pelvic fins either.

Any physically damaged fish, with wounds, spots, lesions, opaque or protruding eyes, torn or ragged fins, should not be considered, neither should a fish that appears inactive. Especially avoid fishes with sunken contours; the knife-edged cross section of a starving fish will never fill out again, even with good feeding.

Until you become experienced, it is a good plan to stick to one dealer. If you 'shop around' you cannot be sure of the origin of any trouble that might manifest itself. With only one source (which might not be to blame at all) you have a single set of evidence to evaluate. You can also more easily strike up a friendly relationship with the dealer, who should not mind dealing with any teething troubles. He should not mind you asking constructive questions either, when you choose your first fishes; it will show him that you are

Above: *A coral island catching site; the first step to captivity for a wide range of tropical marine fishes.*

Below: *In the bag! Recently caught* Amphiprion *and* Abudefduf *species destined for tropical marine tanks.*

47

Above: *Clownfishes (Amphiprion spp.) are kept briefly in well-aerated bowls before being transferred to holding tanks before shipment.*

really taking the task responsibly.

Frequent the shop regularly before actually buying any fishes; make a note of how new arrivals (customers as well as fishes!) are treated. A quiet word with some of the other regular customers will reveal a lot about the dealer's aquarium management techniques – the same faces showing up each week is usually a good sign. Especially find out about whether new fishes are quarantined before being offered for sale.

Other considerations
Always ask about the fishes' feeding habits, and if possible see the fishes actually feeding. A dealer who says

'Oh, they get fed three times a day,' might well be telling the truth, but what you want to know is if the fishes are *eating* three times a day.

It would be a good precaution on your part to find out what S.G. water the fishes are used to; a slight adjustment of your aquarium water conditions beforehand could make all the difference.

Ask if the fishes were caught by means of drugs, as fishes thus caught are less likely to survive in the aquarium and there,is a grave risk to nature conservancy. Fishkeepers would rather, in the main, help to preserve the fishes (and their coral reef homes) in the oceans, rather than have the selfish pleasure of their company for only a few days or weeks, simply because they can be easily and inexpensively captured by using anaesthetizing drugs.

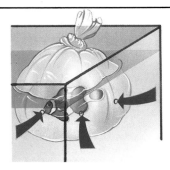 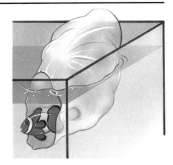

Above: *When introducing new fishes into the aquarium equalize the temperature of the aquarium and transit waters before releasing the fishes. Subdued light lowers stress.*

If you pick a healthy specimen, do not spoil things by catching it with a coarse net (or with a net at all) if you can possibly avoid it. A clear plastic bag or a wide-mouthed jar into which the fish can be herded is a much safer and less damaging way to do it.

Introducing the fishes

The transition from dealer to home aquarium must be made with as little stress to the fishes as possible. They are liable to suffer shock on at least three fronts: temperature, water condition and light.

To avoid stresses from differences in water and temperature conditions, it is normal practice to equalize water temperatures by floating the fishes' transport container in the aquarium, until the two water temperatures are the same. During this period (up to half an hour may be necessary) small amounts of aquarium water can be periodically introduced into the fishes' container so as to acclimatize them gradually to the aquarium's water conditions. Once the temperatures have become equalized, the fish can be released into their new home.

Sudden lighting changes can be very traumatic for fishes, and they should be unpacked and transferred into their new home in subdued lighting conditions.

The above procedures should be followed with only the first purchase of livestock, as the new aquarium can serve as its own quarantine area. Any later additions should be separately

quarantined (see page 50), so that any disease is not introduced into the main aquarium to contaminate healthy fishes.

Aquarium compatibility

It will be a waste of money if you buy fishes that will not tolerate each other's company in your aquarium. Unfortunately, many marine fishes are more than a little anti-social, and strangely enough it is towards fellow members of their own kind that this trait is exhibited most of all.

The following list indicates the degree of compatibility of some of the more popular fishes:

Anemonefishes: Do better with a sea anemone in the tank, and among their own kind.

Angelfishes: Aggressive towards their own kind.

Cardinalfishes: Do not keep with boisterous fishes.

Damselfishes: Gregarious but can quarrel among themselves.

Butterflyfishes: Can be delicate and quarrelsome with like species.

Dragonfishes: Predatory.

Filefishes: Peaceful.

Gobies, Blennies: Can be voracious.

Groupers: Only for large aquariums and never with small fishes.

Seahorses: Better in an aquarium of their own.

Snappers: Not with smaller fishes.

Soldierfishes, Squirrelfishes: Not with smaller fishes.

Surgeons, Tangs: Can be quarrelsome with similar fishes.

Triggerfishes: Not with smaller fishes.

Wrasses: Often perform cleaning duties on other fishes.

Prevention and cure of diseases

It is probably fair to say that, although the average optimistic freshwater fishkeeper never expects disease to strike his tanks, the marine fishkeeper tends to look at life a little more philosophically, accepting that it may be more a case of 'when' rather than 'if'. This is not surprising, when we consider how much more delicate and intolerant of changes marine fishes are in the aquarium.

Many successful marine fishkeepers firmly believe that disease is permanently present in the aquarium (or within the fish), lying dormant and waiting to strike at any fish that falls below par. A common cause that produces fish-weakening conditions is stress; this is brought about by many factors, one of which can be the deviation from the norm of any of the previously described aquarium environmental conditions.

Prevention

It is far better to prevent than to have to cure a disease. You have already made a good start in this direction, by creating the best of conditions in the aquarium and choosing the healthiest-looking fishes.

Once you have established your aquarium, it is vitally important that you allow only healthy additions into it. Your first collection of fishes will have been quarantined in the aquarium itself, during the aquarium's own 'running-in' period. Any new additions must be separately quarantined, and there must be no exceptions to this rule.

Quarantining

It is unlikely that you will buy large quantities of fishes at any one time (the sheer expense will see to that!), and so the quarantine tank need not be a large affair; half the main aquarium size will be suitable.

You may be wondering how functional the quarantine tank should be, for there seems little point in having a second tank permanently furnished waiting for new fishes when perhaps a bare tank with the minimum of 'mod.cons.' will do just as well. Once your marine fishkeeping really becomes established, you will find several uses for that second

furnished tank, in addition to using it as a quarantine area; you may want to rear some young fishes, or you may have to isolate a fish in order to teach it to accept new foods. For a medical treatment centre, however, the fully furnished tank is not entirely suitable: the tank furnishings make accurate dosage of medications difficult; the bacteria in the established biological filter will be harmed by many remedies; and once a treatment has ended (successfully or not) the whole tank should be sterilized and reset up again anyway.

So on balance it seems that a fairly sparsely furnished tank is more practical. An efficient external filter (do not use activated carbon or resin in it when using remedies) and a strong aerator will ensure adequate water treatment. A few flowerpots, or a couple of inclined pieces of slate against the sides, will provide shelter for the fishes. No gravel will be necessary, but the fishes might appreciate it if you paint the base of the tank (outside!) a dark colour. Heating and lighting equipment should be fitted: do not forget the cover glass.

New fish should be carefully introduced into the quarantine tank, and kept there for at least two weeks before they can be considered 'clean'. During this time, take the opportunity to get to know the fish and tempt it to eat. If no disease

Below: *After treating fishes in a separate tank, repeatedly drain off 20% of the water and replace it with new salt water (of the correct S.G.) until all medicated water is replaced.*

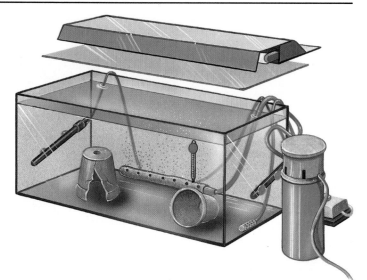

Above: *External filtration is usually adequate for quarantine or treatment tanks as the fishes' stay hopefully will be short. Be sure to provide extra aeration in the tank.*

manifests itself by the end of the quarantine period (many fishkeepers mildly dose the quarantine tank with a proprietary aquarium disinfectant), transfer the fish to the main aquarium, using the same considerations to prevent shock and stress as you did with the first fishes (see page 49). If you have cause to treat the fish while it is in quarantine, you should change back to clean water conditions in the quarantine tank (by gradual partial water changes) before transferring the fish to the main aquarium. Remember to clean out the quarantine tank thoroughly, and disinfect all equipment before using it for a new batch of fishes.

Recognizing diseases

The first clue that things are not quite right in the aquarium is usually some out of the ordinary behaviour of the fishes. (This presumes that you are fully acquainted with their normal demeanour.) Early symptoms of disease are generally fishes being a little 'off-colour' – often a very apt description – and sulking around in a corner; gill movements may become more rapid, and the gills be held wide

open, the fish gasping for breath at the water surface. Swimming actions may become sluggish or erratic. The fish may repeatedly scratch itself on the aquarium decorations, as if attempting to dislodge a persistent irritation. More obvious signs are unnatural spots or blisters to the skin or fins. A particularly distressing symptom (for the fishkeeper to see, as well as for the fish to suffer) is the general wasting away of a fish, even though it is feeding regularly and apparently heartily.

At the onset of any irregular behaviour, you must look first for the obvious and not jump to any rapid conclusions; a fish book can be a little like the family medical dictionary at times, and you may often end up thinking that the fishes have contracted just about everything! Check the number of fishes, the temperature, nitrite, nitrate levels and pH of the water; make sure that the biological filter is working normally and that the external filter's medium does not need renewing.

Once these physical aquarium parameters have been eliminated then a closer study of the fishes is in order. If only one fish seems affected, it should be isolated in the quarantine tank for further observation. If more are affected, then all the fishes should be transferred to the treatment tank for medication: during their absence from the main aquarium, any parasites remaining in it will die through lack of hosts to reinfect. A very good reason for always treating fishes in a separate tank is that many of the remedies, being copper-based, will kill (or otherwise set back) beneficial bacteria, algae and invertebrates (if kept).

Right: *Nature has its own way of dealing with external parasites. Here, a Cleaner Shrimp (*Lysmata *sp.) removes parasites from a willing patient. Fishes will regularly visit a particular coral reef to be cleaned. The Cleaner Wrasse (*Labroides *sp.) also removes parasites.*

DISEASES – Symptoms, Causes and Treatment

SYMPTOMS	CAUSE
Small yellowish-white spots (also on fins). Fishes rub themselves against rocks, stones etc. Eventually gills are affected; respiration becomes difficult	*Oodinium ocellatum*, a parasite
White spots on skin and fins. Symptoms similar to *Oodinium*	*Cryptocaryon irritans*, a parasite; saltwater equivalent of freshwater 'White Spot'
Gills affected, pink rather than red. Mouth kept open. Eyes may become cloudy.	*Benedenia* or *Gyrodactylus*, parasites
Cauliflower-like growth on fins spreading to skin	*Lymphocystis*, virus
Discoloured skin, loss of appetite, open ulcers, vent and fin edges red and inflamed.	*Vibrio anguillarium*, bacteria
Loss of balance	Swim-bladder inflamed and malfunctioning; may be brought on by chilling

Treatment techniques

Never treat the hospital tank with more than one medication at a time. Many remedies may be copper-based and the cumulative effect from several may kill the fishes.

Increase aeration in the treatment tank; remedies often reduce the amount of oxygen available for the fishes. Use only mechanical filtration: bacterial filtration will be adversely affected by the remedies, and activated carbon will reduce the effectiveness of the remedies by adsorbing them from the water.

It may be wise to guard the heater (a wide-mouthed jar or plastic grill could be used) to prevent fish burning

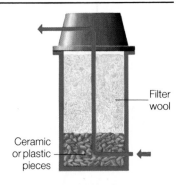

Filter wool

Ceramic or plastic pieces

Above: *Filters serving treatment tanks should have purely mechanical filter media, such as those shown here.*

TREATMENT

Proprietary remedies effective; fish should be quarantined for a further ten days after apparent cure, in case the condition breaks out again from internal infection.

Proprietary remedies effective. Length of treatment may depend on type of remedy used; refer to instructions

Freshwater bath for 15 minutes. Copper-based and anti-parasite remedies also effective.

May be physically cut off or treated locally with tincture of iodine. Often an improvement in conditions will effect a self-cure.

Bath in a general tonic; copper-based remedies and antibiotics also effective. Food soaked in antibiotics helps, as does a clean-up of aquarium conditions.

Bathe fish in warmer water (raise temperature slowly!). Antibiotic-soaked food may also help.

themselves if they attempt to nestle against the heater. Do not allow any contaminated water to be introduced into the main aquarium. Even a wet net previously used in another aquarium can spread disease.

Many symptoms exhibited by fishes are due to incorrect conditions rather than disease or infection. Poisoning may be due to various causes: metal poisoning due to copper or zinc (check that the original tap water supply has not been contaminated through a new plumbing installation and run off several litres to waste before using); overuse of copper-based remedies within the main aquarium (do not do it!); pH, nitrite or nitrate poisoning (check periodically with test kits and make regular partial water changes).

Internal infections are much more problematical, for by the time exterior symptoms are seen it is usually too late to do much about effecting a cure. Make sure that all food given is not contaminated. Good assistance and advice can be obtained from the manufacturers of aquarium products, especially fish food makers, who have had the time and resources to study fish pathology much more thoroughly than the average hobbyist. There are also pathological (and post mortem) services offered by some universities. Should you need to resort to the use of antibiotics, these are generally available only through your veterinarian. A final golden rule – over-medication kills.

Aquarium management

Routine checks

First and foremost will be a 'head count' of the fishes. This is best done at feeding time when the fish should gather together. Any absentee should be located; if it is not in the tank, search the floor – some marine fishes are expert escapers!

It is most important to locate a missing fish in the tank: if it is just not feeling hungry or sociable, that's alright; but if it continues to behave the same way, or is off-colour, then you must assume something is wrong and begin further checks. If the fish is dead, it is vital that you remove the body before decomposition and pollution of the aquarium sets in or the other fishes become infected by disease as a result.

Check the fishes' behaviour; any abnormalities may be an early indication of trouble.

Check regularly the temperature, specific gravity, pH, nitrite and nitrate levels. The pH of the water should be checked very frequently in newly set-up marine aquariums.

Keeping the aquarium healthy

Choose healthy stock and quarantine all new additions. Avoid stressing the fishes in any way.

Don't hope to keep all the fishes that you like the look of; bear in mind

Above: *When checking the sealing, or when cleaning or generally maintaining a very large tank there's only one way to do the job properly! Whatever the size of the tank make sure it is suitable for saltwater use.*

the size of your aquarium and compatibility between the species.

Do not allow any metal objects in the aquarium.

Remove any uneaten food at the earliest opportunity. Acclimatize the fishes to any new food over a period of time – at least several days.

Learn to recognize symptoms of impending water problems: frothy, cloudy, yellowing or smelly water are all signs of deteriorating conditions.

Do not neglect regular partial water changes. Top up evaporation losses with *fresh* water. Make all necessary water condition changes gradual.

Check that water flow rate from filters remains high. Clean external mechanical filters; replace filter medium and activated carbon frequently – at least once a week if necessary. Diminishing water flow from a biological filter may mean one of two things: the substrate may have packed down too tightly (rake it over very lightly), or the airstone in the airlift may be clogged up with salt or calcium (replace the airstone, as boiling it does not always clear the blockage). Empty the collecting chamber of the protein skimmer (if fitted) regularly.

After a period of time, algae will grow all over the tank. This should be removed from the front glass with a non-metal scraper; nylon and plastic scourers are very effective. There is no need to remove algae from the remaining panels of the aquarium, as the fishes will graze on these. Very excessive growths should be thinned out, as a sudden 'algae death' could cause pollution.

Remove any sick fishes from the main aquarium for treatment. Copper-based cures will kill most invertebrates. Do not mix medications. Sterilize all equipment after use and do not use one net between two tanks.

Emergency measures
In the event of heater failure causing a drastic temperature loss, gently re-heat some of the aquarium water in an enamelled pan and return it to the aquarium. Should the electricity supply fail, the large volume of water in the aquarium will act as a heat reserve until the power comes back on again, unless the break in supply is very long. During winter emergencies, the aquarium can be lagged with expanded polystyrene sheets to conserve heat; wrap the aquarium in blankets or layers of newspaper if a serious heat loss looks imminent, and maintain warmth by standing bottles of hot water (heated by alternative means) in the aquarium – look out for overflows.

If the thermostat sticks 'on' then switch off the heating circuit and increase the aeration rate to create more turbulence. Adding cold water directly to the aquarium will cause stress to the fishes. Once the temperature has reached normal levels again, do not forget to switch the heating back on again (having corrected the fault in the meantime).

Look on the bright side
Finally, do not regard any of the above as avoidable chores; most of them should be included in the pleasurable side of fishkeeping, providing for your fishes the care and attention that they undoubtedly deserve. They will repay you in the best manner possible – by presenting you with a living picture of beauty and colour.

Below: *The two impellers in this modern power filter can be seen easily through clear windows once the top has been removed. One of the impellers is shown here out of its housing; inspect and clean regularly as part of essential maintenance.*

Breeding marine fishes

What a wonderful world it would be, if we could continue successful marine fishkeeping into an equally successful breeding programme. Unfortunately, considerably less is known about marine fishes' breeding (both in theory and in practice) than in the well-researched and documented freshwater area.

There is still a great deal of work to be done (and much disappointment to be endured) before even the most easily kept marine fishes can be bred in the aquarium.

To encourage you, the following information may provide some clues which might help to lead you to that first success.

Sexing marine fish is well-nigh impossible, although some authors think that among Angelfishes the male is the larger fish. The best way to obtain a true pair of fishes is to follow the age-old aquarium procedure of having several fishes of the same species together and waiting for natural pairings to occur. The snag here is that you will need a big tank to prevent squabbling between the fishes before true love finds a way!

Spawning attempts will be doomed to failure if the fishes are not in excellent health. They must be fed on the finest foods, preferably with a predominance of live foods and, where necessary, supplementary algae or other green foods.

You should ascertain how the fishes spawn. There are several methods: some species lay eggs on a chosen site, others may be loosely termed egg-scatterers, and others will incubate their eggs in pouches.

Anemonefishes such as Clownfishes and Damselfishes spawn on a site, as do Neon Gobies, and need only a modest-sized aquarium of around 104 litres (23 gallons) capacity. Egg-scattering fishes such as Angelfishes and Butterflyfishes will require a much larger aquarium.

Male Jawfishes and Cardinalfishes incubate their eggs in the mouth, but the male Seahorse hatches the eggs in his abdominal pouch.

The salinity can affect spawning in the aquarium, even though in nature this will not necessarily vary. Reducing the specific gravity (S.G.) slightly – to 1.020, say, or even lower – has been known to trigger off spawning activity in some species of Damselfishes, but you must adjust the S.G. over a period of days so that the fishes are not stressed. If you are lucky and the fishes do spawn, the fry must be raised at this lower S.G. When the adult fishes stop spawning (or the S.G. adjustment ploy does not work), the S.G. must be readjusted back to normal, again gradually over a period of several days.

Below: *Marine fishes spawning in the aquarium is not now such a rare event. Here, a pair of Clownfishes* (Amphiprion clarki) *examine the eggs.*

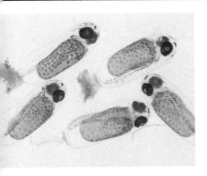

Above: *These 7-day-old embryos (the eyes generally show up clearly first) are of another species of Clownfish,* Amphiprion ocellaris.

Fishes that deposit eggs on a site will usually look after the eggs; the nearest freshwater equivalent to this are the Cichlid fishes. The eggs from egg-scattering species are very numerous, and only a reasonable number (for practical reasons) should be transferred to a hatching tank, which should have the same water conditions as the water in which the fishes spawned.

The hatching aquarium should not have an external power filter, or any filtration system that could trap the young fry; a biological system can be fitted, but should be slow-running. It need not even be operated until the fry are two to three weeks old, as there will not be any significant nitrogenous waste materials until then. If a build-up of ammonia is not checked by the biological filter, regular partial water changes (10%) every other day will be beneficial. An airstone will maintain water circulation, duplicating the water current that pelagic eggs are subjected to in nature.

The biggest problem, after getting the fishes to spawn, will be to raise the fry, and here it is a matter of providing ample, correct-sized food. The highly nutritious brine shrimp nauplius is unfortunately too large to be used as a first food, despite its wide usage in freshwater aquariums for this purpose. Marine rotifers are much preferable, although getting a starter culture may be problematical through the usual sources. It may be advantageous to contact a marine fishkeeping society in this respect.

The following fishes have spawned in captivity, although details are not available as to exact aquarium size conditions. Fry were not necessarily raised successfully in every case.

Angelfish (Egg-scatterer)
Centropyge sp.
Holacanthus sp.
Pomacanthus sp.
Anglerfish (Egg-scatterer)
Antennarius sp.
Butterflyfish (Egg-scatterer)
Chaetodon sp.
Clownfish (Egg-depositor)
Amphiprion sp.
Damselfish (Egg-depositor)
Dascyllus sp.
Abudefduf sp.
Hawkfish (Egg-depositor)
Oxycirrhites sp.
High Hat (Egg-scatterer)
Equetus sp.
Jawfish (Mouth-incubator)
Opistognathus sp.
Jewelfish (Egg-depositor)
Microspathodon sp.
Mandarin Fish (Egg-scatterer)
Synchiropus sp.
Seahorse (Pouch-incubator)
Hippocampus sp.
Wrasse (Egg-scatterer)
Thalassoma sp.

As species to begin with, the Clownfish and the Neon Goby are as good as any, although some American authors prefer to try with the Seahorse, which can often be commercially obtained (in the U.S.A.) as pairs, complete with pregnant female. The theory is that the fry emerge from the male's pouch free-swimming and functional, rather than tiny and extremely delicate fry as born to other species.

As you can see, the above list is not very long when compared with freshwater fishes that have been bred, but it represents no mean achievement by those fortunate enough to have been involved. It also holds out more than just a ray of hope for future marine fishkeepers; so, emboldened by these early triumphs, may you follow their example even more successfully.

Tropical marine fishes

The following pages may not be quite the same as a snorkel-dive beneath the tropical waters over a coral reef, but the photographs illustrate the brilliant colours and fantastic shapes of the fishes that you would indeed find there.

Do not let the pictures tempt you to fill your aquarium (even to the recommended levels) with all and sundry; there are several species that just will not get along together. This is not the fault of the fishes, but of you, the fishkeeper, for presuming that fishes from various areas (with diverse feeding and living habits) will adapt to living together in your aquarium.

The fishes are listed in alphabetical Family order, and within each of these divisions particular groups will again be listed alphabetically. For example, within the Family Chaetodontidae, the fishes are internally classified into Angelfishes, then Butterflyfishes. Similarly,

Pomacentridae is divided into Anemonefishes and Damselfishes.

Sizes given are those accounted for in nature. You should expect some reduction in size for aquarium fishes, although small species generally reach nearer to their mature size than do larger species.

Remember too that you are attempting the almost impossible; the fishes are much more sensitive than the freshwater fishes you may have been used to keeping. Take every possible precaution to acclimatize them correctly, guard them against shock and stress at every turn, feed them carefully, and never neglect any of the aquarium maintenance requirements. Only then will you be repaid by having an aquarium that is a delight to behold, and of which you can be justly proud. Otherwise, you might just as well save your money and perhaps put it towards a trip to see the fishes where they are really at home, on the coral reefs around the world.

Family: ACANTHURIDAE – Surgeonfishes, Tangs and Unicornfishes
The Surgeonfishes, Tangs and Unicornfishes are all oval-shaped and laterally compressed, with a steeply rising forehead, and brightly coloured. But the main characteristic of these fishes is that they have very sharp spines on each side of the caudal peduncle, which can give any other fish a very nasty wound. In some species these spines may be singular and can be erected at will, but in other species two or three fixed spines are present.

Found on coral reefs, mostly in Indo-Pacific waters although a few occur in the tropical Atlantic, they are often intolerant of other fishes (particularly their own kind). However, they are generally herbivorous in their feeding habits. They need a reasonably large aquarium in which there is a natural supply of green foodstuffs such as algae; lettuce or spinach can also be provided.

Acanthurus leucosternon

Powder Blue Surgeonfish
- **Habitat:** Indo-Pacific
- **Length:** 250mm (10in)
- **Diet:** Protein foods and vegetable matter
- **Feeding manner:** Bold grazer
- **Aquarium compatibility:** keep only one in the aquarium

Like all Surgeonfishes, it requires space and well-aerated water.

Below: **Acanthurus leucosternon**
The subtle blue of the body and the bright yellow dorsal fin make this a favourite fish for marine aquariums.

Right: **Naso literatus**
The facial markings would do credit to a beautician. Orange patches near the tail conceal the two scalpels.

Below right: **Zebrasoma veliferum**
The common name of Sailfin Tang can be fully appreciated when the dorsal and anal fins are displayed.

Naso literatus
Japanese Tang; Smooth-headed Unicornfish
- **Habitat:** Indo-Pacific
- **Length:** 500mm (20in)
- **Diet:** Protein foods and greenstuff
- **Feeding manner:** Bold grazer
- **Aquarium compatibility:** Normally peaceful

Has two immovable 'scalpels' on each side of the caudal peduncle.

Zebrasoma veliferum
Sailfin Tang
- **Habitat:** Indo-Pacific, Red Sea
- **Length:** 380mm (15in)
- **Diet:** Protein foods and greenstuff
- **Feeding manner:** Bold grazer
- **Aquarium compatibility:** Normally peaceful, but may be aggressive with large fishes

Young specimens do better in captivity than mature fishes.

61

Family: APOGONIDAE – Cardinalfishes
Cardinalfishes are colourful and carnivorous in nature. They have two dorsal fins, and their large eyes indicate nocturnal activity. They generally frequent coral reefs but are also found in tidal pools and saltwater mangrove swamps. Some species have adopted a ploy, quite common in nature, of 'playing dead' when attacked or frightened.

Above: **Sphaeramia nematopterus**
An instantly recognizable fish, with prominent eyes and two separate dorsal fins. A shy species in the aquarium and nocturnal in habit.

Right: **Balistes vetula**
A very widespread species from the tropical western Atlantic Ocean, known by a host of common names. It grows quickly to a large size.

Sphaeramia (Apogon) nematopterus
Pyjama Cardinalfish
- **Habitat:** Indo-Pacific
- **Length:** 100mm (4in)
- **Diet:** All foods

- **Feeding manner:** Shy
- **Aquarium compatibility:** Do not keep with larger boisterous fishes

Nocturnal by nature; may need acclimatizing with live foods first.

Family: BALISTIDAE – Triggerfishes and Filefishes
Triggerfishes are aptly described: the spiny first dorsal fin can be locked in place when erect by means of a trigger mechanism. Ventral fins, as such, are missing, being replaced by stubby knobs. The first dorsal fin and the primitive ventrals are used to anchor the fish in a crevice and to prevent extraction by predators or fish-collectors. Propulsion is achieved mainly by means of continuous undulating action of the anal and second dorsal fin. Scales may be covered to a smaller or larger extent with tubercles, and in the Filefishes produce a rough skin surface similar to a file. The strong jaws are used to feed on corals, crustaceans and molluscs, and these fishes have also developed special skills for dealing with the Crown-of-Thorns Starfish: they blow it over into an upside-down position and then eat it! Colours can be dazzling, but the Filefishes take things a stage further and are often camouflaged as part of the surrounding scenery, changing their colours almost at will.

Balistes vetula

Queen Triggerfish; Old Wife; Old Wench; Conchino; Peja Puerco
- **Habitat:** Tropical western Atlantic
- **Length:** 500mm (20in)
- **Diet:** Crustaceans, molluscs, small fishes, etc.

- **Feeding manner:** Bold grazer
- **Aquarium compatibility:** Do not keep with small fishes

Tips of the dorsal fins and caudal fin become filamentous with age. Strong jaws with sharp teeth.

Balistoides conspicillum

Clown Triggerfish
- **Habitat:** Indo-Pacific
- **Length:** 500mm (20in)
- **Diet:** Crustaceans, molluscs
- **Feeding manner:** Bold grazer
- **Aquarium compatibility:** Do not keep with small fishes

A very distinctive and easily recognizable species, with its large painted mouth and blotched camouflage colouring.

Left: **Balistoides conspicillum**
In common with other Triggerfishes, this striking species can lock its dorsal spines in a raised position.

Rhinecanthus aculeatus

Picasso Triggerfish
- **Habitat:** Indo-Pacific, Red Sea
- **Length:** 300mm (12in)
- **Diet:** Crustaceans, molluscs, meat foods
- **Feeding manner:** Bold grazer
- **Aquarium compatibility:** Aggressive to the same species and other fishes of the same size

The 'avant garde' colours of this fish make it a popular species.

Below: **Rhinecanthus aculeatus**
Justifiably celebrated for its superb colouring, this splendid species is most peaceable when fairly small.

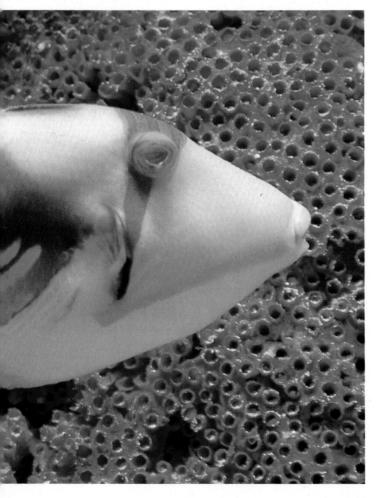

FILEFISHES

Oxymonacanthus longirostris

Beaked Leatherjacket; Orange-green Filefish

- **Habitat:** Indo-Pacific
- **Length:** 100mm (4in)
- **Diet:** Crustaceans, shellfish meat, greenstuff

- **Feeding manner:** Shy grazer, usually in a 'headstanding' manner
- **Aquarium compatibility:** Do not keep with boisterous fishes

Does better with a number of its own species in the aquarium.

Below:
Oxymonacanthus longirostris
Outstanding colour in a small fish.

Top right: **Aspidontus taeniatus**
Do not be fooled by appearances, this is not a true Cleanerfish but an impressive mimic. Keep on its own.

Right: **Ecsenius bicolor**
A beautiful but timid fish that will thrive in a marine aquarium with plenty of retreats. Ideal small size.

Family: BLENNIIDAE – Blennies
Inhabiting rockpools, reefs and shallow waters, Blennies are colourful little characters with long bodies. The eyes are set high on the head, and some species use these when they come out of water, to good effect. Crest-like growths on the head are also common. The dorsal fin is long, almost reaching the caudal, and the ventrals are very small and may be absent altogether in some species. These fishes must be given plenty of hiding places in the aquarium to make them feel entirely at home.

Aspidontus taeniatus
False Cleanerfish
- **Habitat:** Indo-Pacific
- **Length:** 100mm (4in)
- **Diet:** Skin, scales and flesh – preferably from living, unsuspecting victims! In the aquarium they will feed on meat foods, shellfish meat etc.
- **Feeding manner:** Sly
- **Aquarium compatibility:** Do not keep with other fishes

Using its similarity in size, shape and colours to the true Cleanerfish, *Labroides dimidiatus*, this fish approaches its victims who are expecting the usual 'cleaning services'; instead they end up with a very nasty wound and a little bit wiser. Easily distinguished by its underslung mouth; think of a Shark!

Ecsenius (Blennius) bicolor
Two-coloured Slimefish
- **Habitat:** Indo-Pacific
- **Length:** 100mm (4in)
- **Diet:** Meat foods, algae
- **Feeding manner:** Bottom-feeding grazer
- **Aquarium compatibility:** Shy with larger fishes

Provide plenty of hiding places.

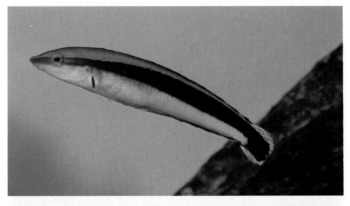

Ophioblennius atlanticus

Atlantic Blenny; Redlip Blenny
- **Habitat:** Tropical western Atlantic
- **Length:** 120mm (4.7in)
- **Diet:** Meat foods, algae
- **Feeding manner:** Bottom-feeding grazer

- **Aquarium compatibility:**
  Territorial, and it chases everything

Keep in a community of small fishes, and provide plenty of hiding places; lengths of plastic pipe placed at the bottom of the tank do nicely.

Family: CALLIONYMIDAE – Dragonets and Mandarinfishes
Dragonets and Mandarinfishes are bottom-dwelling fishes, often with sail-like fins. The eyes are upward-looking. Their colouring is an important aid to camouflaging their existence, and some species also bury themselves in the sand to avoid detection.

Synchiropus picturatus

Psychedelic Fish
- **Habitat:** Pacific
- **Length:** 100mm (4in)
- **Diet:** Small crustaceans, algae
- **Feeding manner:** Shy, bottom-feeding grazer

- **Aquarium compatibility:**
  Intolerant of similarly sized fishes

Best kept in a quiet aquarium away from larger, more lively fishes. It is said that the skin mucus is poisonous, a fact often signalled in gaudy fishes.

Above: **Ophioblennius atlanticus**
Very common off the West Indies, this distinctive species has the typical high forehead and crestlike growths.

Below: **Synchiropus picturatus**
A dazzling subject for the aquarium. The similar Synchiropus splendidus (Mandarinfish) is more widely seen.

Family: CHAETODONTIDAE – Butterflyfishes and Angelfishes

This Family provides the marine aquarium with some of the most attractive specimens. Sizes range from the 200mm (8in) of the Butterflyfishes to around 600mm (24in) of the much larger Angelfishes. The thicker-bodied Angelfishes may be further distinguished from the Butterflyfishes by the presence of a spine at the base of the gill cover.

The fishes are laterally compressed and high-bodied, some with very 'snouty' mouths, which they use to extract food from crevices among the coral. Their amazing colour patterns serve to camouflage and protect vulnerable parts of their bodies, and to assist species identification. Juvenile Angelfishes have a different colouration pattern from that of the adult. It is not always possible to identify positively a young Angelfish; many are similarly marked in white on a blue background, presumably an effective camouflage pattern.

Butterflyfishes and Angelfishes are not beginners' fishes, by any means. They are easily upset by changes in water conditions, and usually show any dissatisfaction with aquarium life by going on hunger-strike: one day they are quite content with the diet provided, the next day they will not touch it. Many are algae-eaters and nibblers at coral heads; living corals and sea anemones will not last long with these fishes in the same aquarium.,

They are fairly territorial, and intolerant of their own kind (particularly the Angelfishes), which rather limits the number that you can keep together; but in a large aquarium with a good number of retreats you can expect better results.

ANGELFISHES

Centropyge argi

Pygmy Angelfish; Cherubfish
- **Habitat:** Western Atlantic
- **Length:** 75mm (3in)
- **Diet:** Meat foods and plenty of greenstuff
- **Feeding manner:** Grazer
- **Aquarium compatibility:** Can be aggressive; more peaceable if kept in pairs.

A deeper water fish, which lives around the base of the reef rather than at the top. Colour pattern details around the head may vary between species, and there is no different juvenile colour form as in other species of Angelfishes.

Other species:

Centropyge bicolor

Bicolor Cherub; Oriole Angel; Black and Gold Angelfish
- **Habitat:** Pacific
- **Length:** 125mm (5in)

Above right: **Centropyge argi**
This Angelfish may lose out in the race for food at feeding times if kept in a tank with larger fishes.

Right: **Centropyge bicolor**
An attractive and compact Angelfish with a bold marking on the head.

Centropyge loriculus
Flame Angelfish
- **Habitat:** Pacific
- **Length:** 100mm (4in)

Euxiphipops xanthometapon
Yellow-faced Angelfish
- **Habitat:** Indo-Pacific
- **Length:** 380mm (15in)
- **Diet:** Meat foods and greenstuff
- **Feeding manner:** Grazer
- **Aquarium compatibility:** Only young specimens are suitable

For the experienced fishkeeper only.

Above: **Centropyge loriculus**
A choice member of the Centropyge genus that makes up in vivid coloration what it lacks in size.

Holacanthus ciliaris
Queen Angelfish
- **Habitat:** Western Atlantic
- **Length:** 460mm (18in)
- **Diet:** Meat foods and greenstuff
- **Feeding manner:** Grazer
- **Aquarium compatibility:** Grows large; young specimens are ideal

A very beautiful fish in the aquarium. Variations in colour patterns occur.

Left: **Holacanthus ciliaris**
*It is easy to see why this stunning
Queen Angelfish is classed by most
marine fishkeepers as the fish they
would most love to keep.*

Above:
Euxiphipops xanthometapon
*This Yellow-faced Angelfish seems to
be arrogantly demanding homage to
its regal appearance. Needs care.*

Pomacanthus paru
French Angelfish
- **Habitat:** Western Atlantic
- **Length:** 300mm (12in)
- **Diet:** Meat food and greenstuff
- **Feeding manner:** Grazer
- **Aquarium compatibility:**
  Aggressive and can be bullies

Juvenile members of the Atlantic Pomacanthids exhibit cleaning tendencies towards other fishes, and each Angelfish's territory is recognized as a cleaning station. Young fishes are black and marked with yellow vertical bands.

Below: **Pomacanthus paru**
A Cleanerfish turns away from this French Angelfish, unperturbed by the vast difference in their sizes.

Pomacanthus semicirculatus
Koran Angelfish
- **Habitat:** Indo-Pacific, Red Sea
- **Length:** 400mm (16in)
- **Diet:** Meat food and greenstuff
- **Feeding manner:** Grazer
- **Aquarium compatibility:**
  Territorial

Juvenile white markings form semi-circles rather than straight lines; markings on the caudal fin during the colour change to adulthood often resemble Arabic script in the Koran.

Right: **Pomacanthus paru**
Similar in shape to the adult, this juvenile has a quite different pattern of markings on its body.

Above:
Pomacanthus semicirculatus
The juveniles of many Angelfishes are dissimilar in colour and patterning to the adult fishes. This makes buying young Angelfishes something of a lottery. This juvenile Koran Angelfish has elegant curved markings.

Below:
Pomacanthus semicirculatus
The fully mature Koran Angelfish loses its bright blue coloration but still retains some markings on the head and on the caudal fin.

Above: **Chaetodon auriga**
Hardy in an aquarium, this is one of the most attractive and popular of all the Butterflyfishes usually imported.

Below: **Chaetodon chrysurus**
A beautiful but variable species that deserves to be more widely available than it is. Shy in the aquarium.

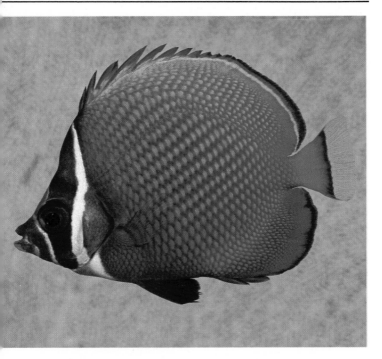

BUTTERFLYFISHES

Chaetodon auriga
Threadfin Butterflyfish
- ● **Habitat:** Indo-Pacific, Red Sea
- ● **Length:** 200mm (8in)
- ● **Diet:** Crustaceans, polyps, algae
- ● **Feeding manner:** Grazer
- ● **Aquarium compatibility:**
  Peaceful but shy

The common name refers to a threadlike extension to the dorsal fin.

Chaetodon chrysurus
Pearlscale Butterflyfish
- ● **Habitat:** Indo-Pacific, Red Sea
- ● **Length:** 150mm (6in)
- ● **Diet:** Crustaceans, vegetable matter
- ● **Feeding manner:** Grazer
- ● **Aquarium compatibility:** A shy species

The scales on this species are large and dark-edged, giving the fish a lattice-covered or checkered pattern. The main feature is the vertical arc-shaped band of orange connecting

Above: **Chaetodon collare**
Easy to recognize, due to its brown colour and reticulated patterning. One for experienced fishkeepers.

the rear of the dorsal and anal fins; a repeated orange band appears in the caudal fin. The fish's habitat is thought to be nearer to Africa, Mauritius and the Seychelles rather than spread widely over the Indo-Pacific area. Not a commonly imported fish for marine aquariums.

Chaetodon collare
Pakistani Butterflyfish
- ● **Habitat:** Indian Ocean
- ● **Length:** 150mm (6in)
- ● **Diet:** Meat food and greenstuff
- ● **Feeding manner:** Grazer
- ● **Aquarium compatibility:** May be intolerant of other members of its own, or other, species

Unusual coloration for a Butterflyfish. Reputedly difficult to keep in an aquarium, although not all authorities seem to agree on this. Not suggested as a beginner's fish.

Chaetodon ephippium
Saddleback Butterflyfish
- **Habitat:** Indo-Pacific
- **Length:** 230mm (9in)
- **Diet:** Crustaceans
- **Feeding manner:** Grazer
- **Aquarium compatibility:** May be intolerant of other members of its own, or other, species

May be difficult to acclimatize to a successful aquarium feeding pattern.

Chaetodon semilarvatus
Addis Butterflyfish
- **Habitat:** Indian Ocean, Red Sea
- **Length:** 200mm (8in)
- **Diet:** Crustaceans, coral polyps, algae
- **Feeding manner:** Grazer
- **Aquarium compatibility:** May be intolerant of other members of its own, or other, species

Adapts fairly readily to aquarium life.

Below: **Chaetodon semilarvatus**
Two lovely blue eyes relieve the bright yellow body of the Addis Butterflyfish. This is a typical member of the Family. The pectoral fins provide forward propulsion; the caudal fin steers.

Above: **Chaetodon ephippium**
Strikingly patterned but not easy.

Right: **Chaetodon striatus**
An adult photographed at rest during the night on a Florida coral reef.

Chaetodon striatus
Banded Butterflyfish
- **Habitat:** Tropical Atlantic Ocean
- **Length:** 150mm (6in)
- **Diet:** Crustaceans, coral polyps, algae
- **Feeding manner:** Grazer
- **Aquarium compatibility:** May be intolerant of other members of its own, or other, species

An easily recognizable species, with four dark bands across the body. A continuous dark band passes through the outer edges of the dorsal, caudal and anal fins to connect with the ends of the third vertical band. The juvenile form has a white-ringed black spot on the soft dorsal fin. A good community fish but prone to stress from any kind of shock.

Chelmon rostratus

Copper-banded Butterflyfish
- **Habitat:** Indo-pacific, Red Sea
- **Length:** 170mm (6.7in)
- **Diet:** Small animal foods, algae
- **Feeding manner:** Picks at coral heads
- **Aquarium compatibility:** Aggressive to its own species

May take time to acclimatize to aquarium foods; very sensitive to deteriorating water conditions.

Above: **Chelmon rostratus**
This Butterflyfish has evolved further than the usual body shape and has developed a long snout ideal for feeding from coral crevices.

Below: **Heniochus acuminatus**
Often mistaken for the Moorish Idol, this Butterflyfish has a much longer dorsal fin (hence Pennant Coralfish) but lacks the colours around the face.

Above: **Forcipiger longirostris**
Strongly resembling the Copper-banded Butterflyfish in body shape, this species has a similar predilection for searching crevices for food.

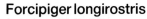

Forcipiger longirostris
Long-nosed Butterflyfish
- **Habitat:** Indo-Pacific, Red Sea
- **Length:** 200mm (8in)
- **Diet:** Small animal foods, algae
- **Feeding manner:** Picks at coral heads
- **Aquarium compatibility:** Not so aggressive as the Copper-banded Butterflyfish

May take some time to acclimatize to aquarium foods; very sensitive to deteriorating water conditions.

Heniochus acuminatus
Wimplefish; Pennant Coralfish; Poor Man's Moorish Idol
- **Habitat:** Indo-Pacific, Red Sea
- **Length:** 200mm (8in)
- **Diet:** Meat food and greenstuff
- **Feeding manner:** Grazer
- **Aquarium compatibility:** Not so territorial as other Butterflyfishes

Young specimens frequently act as cleanerfishes. Appreciates plenty of swimming room in the aquarium.

Family: CIRRHITIDAE – Hawkfishes

Hawkfishes are streamlined pencil-shaped fishes that like to perch on outcrops of rocks or coral. They feed on small live foods which they capture by short, sharp dashes from their usual perching place. Many are nocturnal.

Oxycirrhites typus

Longnosed Hawkfish
- ● **Habitat:** Indian Ocean mainly
- ● **Length:** 100mm (4in)
- ● **Diet:** Meat food
- ● **Feeding manner:** Sits on a rock or coral, then dashes out to grab food
- ● **Aquarium compatibility:** Peaceful

Generally hardy in the aquarium.

Right: **Chilomycterus schoepfi**
Two prickly-looking specimens. Smaller than other related species.

Below right: **Diodon hystrix**
A young specimen, seen against a backdrop of Red Organ-pipe Coral.

Below: **Oxycirrhites typus**
A typical pose, sitting at the 'crossroads' of the aquarum waiting for anything edible to pass by.

Family: DIODONTIDAE – Porcupinefishes

Similar in form to the Pufferfishes, members of the Diodontidae Family can be distinguished by their teeth, which are fused together to form the equivalent of a beak (more like a tortoise's jaw than a bird's beak). Hard-shelled foods (mussels, shrimps, etc) present no problems to these fishes.

The body spines are usually held flat but they share the ability of Pufferfishes to inflate themselves, which causes the spines to stand out from the body in a threatening manner to dissuade the approach of would-be predators.

Chilomycterus schoepfi

Spiny Boxfish
- ● **Habitat:** Tropical Atlantic
- ● **Length:** 250mm (10in)
- ● **Diet:** Crustaceans, molluscs, meat food
- ● **Feeding manner:** Bold
- ● **Aquarium compatibility:** Aggressive among themselves; do not keep them with small fishes

Spines are fixed and held more erect. Not a frequent 'inflater' but it does make croaking sounds out of water.

Diodon hystrix

Common Porcupinefish
- ● **Habitat:** All tropical seas
- ● **Length:** 900mm (36in). It is considerably smaller in the aquarium
- ● **Diet:** Crustaceans, molluscs, meat food
- ● **Feeding manner:** Bold
- ● **Aquarium compatibility:** Generally peaceful to other fishes

Spines held flat against the body. Can become tame enough to hand-feed.

Family: LABRIDAE – Wrasses and Rainbowfishes

This large Family contains many colourful fishes whose juvenile forms are quite suitable for the aquarium. The Wrasses and Rainbowfishes are also interesting on several fronts: they have juvenile colouration patterns that are different from those of the adult fish; many bury themselves at night or spin mucus cocoons in which to rest; and a number of species perform 'cleaning services' on other species, removing parasites in the process. Most enjoy molluscs and crustaceans, but they also will eat green food.

Coris gaimardi

Clown Wrasse; Red Labrid
- **Habitat:** Indo-Pacific
- **Length:** 400mm (16in)
- **Diet:** Crustaceans, shellfish meat
- **Feeding manner:** Bold bottom-feeder
- **Aquarium compatibility:** May be quarrelsome among themselves

The colour patterns of juvenile and adult are quite different. May be nervous, and dash about when first introduced; try not to shock them.

Right: **Coris gaimardi**
The coloration patterns of juvenile species of the genus Coris *are quite unlike those found in the adults. This juvenile has markings similar to the Clown Anemonefish (*Amphiprion sp*). This species is often confused with* Coris formosa, *which is similarly patterned but with white markings extending further down the sides.*

Below: **Coris gaimardi**
The adult fish is still strikingly coloured, but in a different way.

84

Labroides dimidiatus

Cleaner Wrasse

- ● **Habitat:** Indo-Pacific
- ● **Length:** 100mm (4in)
- ● **Diet:** Skin parasites of other fishes, in nature; in captivity, finely chopped meat foods make an excellent substitute
- ● **Feeding manner:** Bold
- ● **Aquarium compatibility:** Peaceful

A valuable asset in the aquarium. Beware of the predatory look-alike *Aspidontus taeniatus* (see page 67).

Right and below:
Labroides dimidiatus
The true Cleaner Wrasse has its mouth situated at the front tip of the head. Most fishes remain quite still while they are being cleaned.

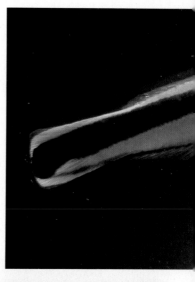

Thalassoma lunare
Green Parrot Wrasse
- **Habitat:** Indo-Pacific
- **Length:** 330mm (13in)
- **Diet:** All foods
- **Feeding manner:** Bottom-feeder
- **Aquarium compatibility:**
  Peaceful, but its constant
  movement may disturb smaller
  fishes. Needs plenty of room

Adult specimens lose the dark
blotches of the juveniles and are
bright green with red and blue
patterns on the head.

Below: **Thalassoma lunare**
*This is quite a hardy species and easy
to keep, although it can be irritatingly
overactive for smaller fishes. Provide
a spacious tank to accommodate its
abundant energy.*

Above: **Opisthognathus aurifrons**
*Although Yellow-headed Jawfish is a
suitable name for this fish, perhaps
'Jack-in-the-Box' might be more
accurate, as the fish remains quite still
waiting for small live food to pass by
before jumping out to take it.*

Above right: **Lactoria cornuta**
*It might seem that the rigid exterior
covering of the Long-horned Cowfish
would make swimming difficult. Very
precise movements are achieved,
however, by careful use of the dorsal,
anal and pectoral fins.*

Family: OPISTHOGNATHIDAE – Jawfishes
The Jawfishes are burrowing fishes that live in holes in the seabed, entering tail-first with a fine turn of speed if danger threatens. They are not very adventurous, hovering vertically in their holes waiting for small live foods to pass by. Some species are solitary; others live in social colonies.

Opisthognathus aurifrons
Yellow-headed Jawfish
- **Habitat:** Western Atlantic
- **Length:** 125mm (5in)
- **Diet:** Finely-chopped shellfish meat
- **Feeding manner:** Makes rapid lunges from a vertical hovering position near its burrow to grab any passing food

- **Aquarium compatibility:**
  Peaceful, and may not even be bothered by other fishes, although the Royal Gramma (*Gramma loreto*) may try to steal its territory

Will excavate a burrow in which to live, entering the hole tail-first at any sign of trouble. Very delicately coloured. Good jumper.

Family: OSTRACIONTIDAE – Boxfishes, Cowfishes and Trunkfishes
The Boxfishes, Cowfishes and Trunkfishes in this group have a hard, shell-like exterior; Cleanerfishes may damage their sensitive skin. The ventral fins are absent. It seems that most growth occurs in the caudal peduncle area rearwards. They are slow-moving (some have been designated 'Hovercraft Fishes' by imaginative authors) and they do indeed have a similar form of locomotion. They have the unfortunate ability to excrete a toxic substance when frightened (it kills them too!) or killed, which makes them a slightly risky proposition for a community aquarium.

Lactoria cornuta
Long-horned Cowfish
- **Habitat:** Indo-Pacific
- **Length:** 500mm (20in)
- **Diet:** Crustaceans and greenstuff
- **Feeding manner:** Shy bottom-feeder with a browsing action

- **Aquarium compatibility:**
  Intolerant towards each other

The two 'horns' on the head and two more projections at the bottom rear of the body make for easy identification. The body often has blue spots.

Left: Ostracion meleagris
*A female; the male is so different
(lower yellow-dotted purple area,
white-dotted top) that it is often,
wrongly, called* O. lentiginosum.

Below: **Plotosus lineatus**
*The only Catfish in the marine
aquarium world. The spines on the
dorsal and pectoral fins are
poisonous, so handle it with care.*

Family: PLOTOSIDAE – Marine Catfishes
These marine Catfishes are very smart-looking when young but the elegant
stripes, together with the strong shoaling tendency, are lost with age.

Ostracion meleagris
Blue-spotted Boxfish
- **Habitat:** Indo-Pacific
- **Length:** 200mm (8in)
- **Diet:** Crustaceans
- **Feeding manner:** Bottom-feeder
- **Aquarium compatibility:**
  Peaceful

Start feeding with brine shrimp,
Daphnia etc. Not easy to keep.

Plotosus lineatus
Saltwater Catfish
- **Habitat:** Indo-Pacific
- **Length:** 300mm (12in)
- **Diet:** Chopped shellfish meat
- **Feeding manner:** Bottom-feeder
- **Aquarium compatibility:**
  Peaceful

Young specimens shoal together
when threatened and form a ball-
like clump with the heads outwards.

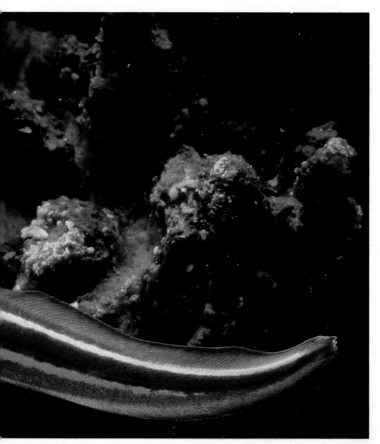

Family: POMACENTRIDAE – Anemonefishes (Clownfishes) and Damselfishes
The Anemonefishes and Damselfishes in this Family are among the fishes commonly kept by the newcomer to marine fishkeeping. All are highly coloured, lively and hardy. Each has a distinctive swimming style: the Anemonefishes seem to waddle, and the Damselfishes have a more bouncy action as if they were constantly being affected by the waves above their heads.

As their name suggests, the Anemonefishes are never found too far away from the sea anemones with which they share a symbiotic relationship. The Anemonefish was thought to be immune to the stinging cells of the sea anemone, but it now appears that the mucus on the fish prevents the stinging cells from being activated.

Despite their natural lifestyle, Anemonefishes will live quite happily in an aquarium without the presence of a sea anemone, but then you will not see the fish behave as it does in the wild. Large sea anemones can 'house' several Anemonefishes, but the fish will become territorial in respect of a smaller, 'owner-occupied' sea anemone. Spawning has occurred in the aquarium, and the eggs are laid on a selected site and guarded.

Damselfishes are more than happy using the coral heads on the reef as their territory, and are among the most agile of fishes when you are trying to catch them. Some are quarrelsome with their own kind, although young specimens will often school together. Widely distributed in all warm seas.

Below: **Amphiprion ocellaris**
Keep these attractive fishes in a small shoal and, ideally, together with a sea anemone in the aquarium. A perfect choice for all new to fishkeeping.

Right: **Amphiprion nigripes**
More subtle in colour and patterning than the rather exaggerated markings of some Clownfishes. A handsome and reliable aquarium subject.

ANEMONEFISHES

Amphiprion nigripes
Black-footed Clownfish
- **Habitat:** Indian Ocean
- **Length:** 80mm (3.2in)
- **Diet:** Plankton and crustaceans in nature, but finely chopped foods are ideal in captivity
- **Feeding manner:** Bold feeder
- **Aquarium compatibility:** Best kept with other Anemonefishes

If you keep sea anemones in the tank for the benefit of the Anemonefishes, you must exclude fishes that would eat them. A lively and hardy species.

Amphiprion ocellaris
Orange Clownfish
- **Habitat:** Indo-Pacific
- **Length:** 80mm (3.2in)
- **Diet:** Finely chopped food
- **Feeding manner:** Bold
- **Aquarium compatibility:** Sometimes territorial to the exclusion of other Anemonefishes.

Often confused with a very similar species, *Amphiprion percula*.

93

Premnas biaculeatus

Maroon Clownfish; Spine-cheeked Anemonefish; Tomato Clownfish
- **Habitat:** Pacific Ocean
- **Length:** 150mm (6in)
- **Diet:** Finely chopped food
- **Feeding manner:** Bold
- **Aquarium compatibility:** Aggressive with other Anemonefishes

Has two spines on each side of the head, one beneath the eye and another on the gill cover.

Right: **Premnas biaculeatus**
Nearly twice the size of other Clownfishes, this species has spines on the head and on the gill cover.

Below right: **Chromis caerulea**
Many of the Chromis genera have iridescent colours, which suits their vivacious nature perfectly.

Below: **Abudefduf saxatilis**
A very common, hardy Damselfish; it seems to be an ideal fish with which to help condition any new tank.

DAMSELFISHES

Abudefduf saxatilis

Sergeant Major
- **Habitat:** Indo-Pacific, tropical Atlantic
- **Length:** 175mm (6.8in)
- **Diet:** Finely chopped meat, algae and greenstuff
- **Feeding manner:** Bold grazer
- **Aquarium compatibility:** Juveniles very active; adults may become aggressive

A hardy fish, and an ideal beginner's choice. A shoal in a large aquarium makes an impressive sight.

Chromis caerulea

Green Chromis; Blue Puller
- **Habitat:** Indo-Pacific, Red Sea
- **Length:** 120mm (4.7in)
- **Diet:** Chopped meat
- **Feeding manner:** Bold
- **Aquarium compatibility:** May be a fin-nipper

A hardy, colourful shoaling species.

Chromis cyanea
Reef Fish
- **Habitat:** Tropical Atlantic
- **Length:** 175mm (6.8in)
- **Diet:** Chopped meat
- **Feeding manner:** Bold
- **Aquarium compatibility:**
  Peaceable shoaling fish that prefers to be with some of its own kind to feel completely at home.

Likes vigorously aerated water to simulate its native reef conditions.

Above: **Chromis cyanea**
It is easy to visualize a shoal of these superb blue Reef Fishes swimming around the coral heads.

Below right: **Dascyllus marginatus**
A boldly marked, active reef fish for the tropical marine aquarium.

Below: **Dascyllus trimaculatus**
The popular Domino Damsel positively bounces with vitality, but it is very possessive about its territory.

Dascyllus marginatus
Marginate Damselfish;
Marginate Puller
- **Habitat:** Red Sea
- **Length:** 100mm (4in)
- **Diet:** Chopped meat
- **Feeding manner:** Bold
- **Aquarium compatibility:** Can be territorially minded

In common with all *Dascyllus* species, this fish occasionally makes audible purring or clicking sounds.

Dascyllus trimaculatus
Domino Damsel; Three Spot
Humbug; White Spot Puller
- **Habitat:** Indo-Pacific, Red Sea
- **Length:** 125mm (5in)
- **Diet:** Chopped meat and even dried foods
- **Feeding manner:** Bold
- **Aquarium compatibility:** Territorial

The white spots may fade with age. Lively and adaptable aquarium fish.

Paraglyphodon melanopus

Yellow-backed Damselfish
- **Habitat:** Indo-Pacific
- **Length:** 75mm (3in)
- **Diet:** Chopped meat
- **Feeding manner:** Bold
- **Aquarium compatibility:** May be aggressive towards its own and smaller species

A spacious tank with plenty of hiding places suits this species very well.

Below: **Paraglyphodon melanopus**
A striking species, also known as the Bowtie, or Blue Fin, Damsel.

Pomacentrus coeruleus

Blue Devil; Electric-blue Damsel
- **Habitat:** Indo-Pacific
- **Length:** 100mm (4in)
- **Diet:** Chopped meat
- **Feeding manner:** Bold
- **Aquarium compatibility:** Pugnacious

A hardy species in the marine aquarium.

Right: **Pomacentrus coeruleus**
Although very attractive in the aquarium, this species unfortunately lives up to its name of Blue Devil by being rather pugnacious and territorial, as any intruders discover.

Left: **Pterois volitans**
A popular aquarium fish. There is no tissue between the pectoral fin rays.

Above: **Dendrochirus zebra**
Hardy. There is a thin layer of tissue between the pectoral fin rays.

Family: SCORPAENIDAE – Dragonfishes, Lionfishes, Scorpionfishes and Turkeyfishes

Here are the exotic 'villains' of the aquarium. They are predatory carnivores that glide up to their prey and then engulf them with their large mouths. The very ornamental fins are not just for looks, either; they have poisonous stinging cells and will inflict a very painful wound, so handle these species with care.
Obviously these fishes require some form of live food – usually Goldfishes – if they are to thrive in captivity, although some species will accept chopped meat.

Dendrochirus zebra
Dwarf Lionfish
- **Habitat:** Indo-Pacific
- **Length:** 200mm (8in)
- **Diet:** Small fishes, meat chunks and live Guppies
- **Feeding manner:** Slow-swimming sudden gulper
- **Aquarium compatibility:** Distinctly unsociable

Pterois volitans
Dragonfish; Red Firefish
- **Habitat:** Indo-Pacific
- **Length:** 350mm (13.8in)
- **Diet:** Smaller fishes
- **Feeding manner:** Slow-swimming sudden gulper
- **Aquarium compatibility:** Unsociable. Keep on its own in the aquarium or with larger fishes.

Family: SERRANIDAE – Sea Basses

This large Family of predatory fishes, the Sea Basses, has many members whose juvenile forms have become aquarium favourites. Equally popular are the Basslets, whose brilliant colours make them popular among fishkeepers. The majority of these species need a large aquarium, and their diet must include crustaceans and meaty foods. The two species described on this page are often classified in separate groups, the Anthiidae and Grammidae, but are included together here as a sub-group of the Serranidae for convenience.

Anthias squamipinnis
Wreckfish; Orange Sea Perch;
Lyre-tail Coralfish
- **Habitat:** Indo-Pacific
- **Length:** 125mm (5in)
- **Diet:** Live foods preferred, then meat foods
- **Feeding manner:** Bold, but prefers moving foods
- **Aquarium compatibility:** Generally peaceful. Likes companions of the same species

Very beautiful. The male can be distinguished from the female by its elongated rays in the dorsal fin.

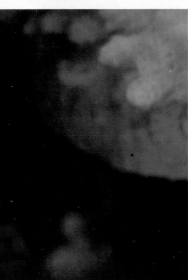

Above: **Anthias squamipinnis**
A striking fish that swims with a distinctive lazy undulating action.

Left: **Gramma loreto**
A secretive cave dweller with quite exquisite pattern and colouring.

Gramma loreto
Royal Gramma; Fairy Basslet
- **Habitat:** Western Atlantic
- **Length:** 130mm (5in)
- **Diet:** Finely chopped meat foods, and brine shrimp
- **Feeding manner:** Rather shy
- **Aquarium compatibility:** Aggressive to its own kind. Do not keep with boisterous species.

Remarkable colouring. An almost identical species, *Pseudochromis paccagnellae*, has a narrow white line dividing the two main body colours.

Family: SIGANIDAE – Rabbitfishes

These fishes are herbivorous but they should be handled with care, for they have sharp spines on their fins. The Foxface, or Foxfish, differs from the normal rabbit-mouthed appearance of this Family, in that its snout is tubular.

Lo (Siganus) vulpinus
Foxface; Foxfish; Badgerfish
- **Habitat:** Pacific
- **Length:** 250mm (10in)
- **Diet:** Live food, meat foods and greenstuff
- **Feeding manner:** Bold grazer adopting a typical a 'head-down' feeding attitude

Below: **Lo (Siganus) vulpinus**
This fish hovers around crevices, raising its spines when threatened.

- **Aquarium compatibility:** Lively but peaceable

Although physically similar to the Surgeonfishes (Acanthuridae), it has no spines on the caudal peduncle.

Right: **Hippocampus kuda**
Seahorses are a major attraction to marine fishkeepers. This newly imported Yellow Seahorse should revert to a light yellow colour as it settles down into its new home.

Family: SYNGNATHIDAE – Pipefishes and Seahorses

The extremely engaging, and immediately recognizable, Seahorses are the fishes that everyone wants to keep in the aquarium. They are poor swimmers, and require considerable quantities of small live foods. They make their home among coral heads, where they attach themselves by their prehensile tails. Their breeding method is interesting: the fertilized eggs are incubated in the male's abdominal pouch for four to five weeks before they hatch.

Hippocampus kuda
Yellow Seahorse
- **Habitat:** Indo-Pacific
- **Length:** 250mm (10in) – vertically!
- **Diet:** Plenty of live foods – young freshwater live-bearer fry, very small crustaceans, *Daphnia*, etc.
- **Feeding manner:** Shy

- **Aquarium compatibility:** Best kept in a species tank

An irresistible fish, with a very unusual method of reproduction. Needs anchorage points in the aquarium, so provide plenty of suitably branched decorations for the prehensile tails.

Family: TETRAODONTIDAE – Pufferfishes
Pufferfishes lack the spines of their near relatives the Porcupinefishes
(Diodontidae); and their jaws, although fused, are divided into two parts.
(Tetraodon = Four-toothed, unlike Diodon = two-toothed). Ventral fins are
absent. They have poisonous flesh. Species of *Tetraodon* (among which are
some freshwater members) are fully inflatable, but members of the genus
Canthigaster are able to inflate themselves only partly to deter predators.

Canthigaster solandri

Sharp-nosed Puffer
- **Habitat:** Indo-Pacific, Red Sea
- **Length:** 120mm (4.7in)
- **Diet:** Finely chopped meat food
- **Feeding manner:** Bold grazer.
  Browses on live foods on coral

- **Aquarium compatibility:**
  Generally peaceful in the aquarium
  except to members of its own kind

Swims with caudal fin folded. Often
classified in a separate Family, the
Canthigasteridae.

Family: ZANCLIDAE
The Toby is sometimes classified in the Family Acanthuridae because of the
physical form of the young fish, although it is difficult to see any resemblance at
first glance in the adult; others feel that it is superficially nearer to the
Chaetodontidae Family, especially the genus *Heniochus*. The common name
of Moorish Idol is derived from the high esteem in which the fish is held by some
Moslem populations. Adult fishes have 'horns' in front of the eyes.

Zanclus cornutus

Moorish Idol; Toby
- **Habitat:** Indo-Pacific
- **Length:** 250mm (10in)
- **Diet:** Small crustaceans, chopped
  meat foods and greenstuff
- **Feeding manner:** Bold grazer in
  nature, but shy in the company of
  other fishes

- **Aquarium compatibility:** A
  sensitive shoaling fish that is
  difficult to acclimatize to aquarium
  life, especially feeding

Moorish Idols may quarrel among
themselves in the close confines of
the aquarium. Intolerant of disease
remedies. Not a beginner's fish.

Above: **Zanclus cornutus**
Plentiful in nature, congregating in shoals, but difficult to keep with great success in captivity.

Below: **Canthigaster solandri**
An attractive fish that lies on the seabed at night. It is also known as Canthigaster margaritatus.

Coldwater marine fishes

Most fishkeepers are attracted by any aquatic life, regardless of their particular individual interests, and a visit to the seashore provides an opportunity to explore rockpools and their intriguing inhabitants.

If you are going to the seashore in order to collect species, you must be well-equipped to transport any livestock that you capture. Large plastic buckets with clip-on lids are ideal, although a double thickness of 'cling-film' may be an adequate substitute for a lid. A battery-operated air pump will ensure that the livestock survive any lengthy journey home better, especially during the summer months in one of those interminable traffic jams. Take extra buckets or other suitable sealable containers, in which to collect some *unpolluted* sea water.

Collect specimens with care, and remember to leave the rockpool in a fit state for the animals you leave behind, not as if a bomb had hit it. If you collect invertebrates such as sea anemones or starfishes, do not attempt to dislodge them from their chosen site; collect site and animal together, replacing any rocks removed with others, to restore the number of hiding places in the pool.

Family: BLENNIIDAE – Blennies

These are most common in rockpools, and are found hiding under overhanging rocks. They are often confused with Gobies, but they lack the 'suction cup' formed by the fusion of the pectoral fins. Most Blennies have a tentacle over each eye. They are easy to keep but rather drab in colour.

Blennius gattorugine

Tompot Blenny
- **Habitat:** Mediterranean, Eastern Atlantic from West Africa to Scotland
- **Length:** 200mm (8in)

Left: **Blennius gattorugine**
A typical pose – viewing the world from a secure refuge beneath a rock.

- **Diet:** Live foods, meat and worms
- **Feeding manner:** Bottom-feeder
- **Aquarium compatibility:** Can be territorial; may worry smaller fishes – and be themselves worried by larger ones

They prefer a tank decorated with medum-sized stones under which they can hide. May become tame.

Family: GOBIIDAE – Gobies

Gobies have no lateral line system along the flanks of the body; instead, they have sensory pores (connected to the nervous system) on the head and over the body. They can live for quite a long time – up to ten years has been recorded. The ventral fins are fused into a 'suction cap'

Gobius cruentatus

Red-mouthed Goby
- **Habitat:** Eastern Atlantic, from North Africa to Southern Ireland
- **Length:** 180mm (7in)

Left: **Gobius cruentatus**
This Red-mouthed Goby's colour pattern camouflages it equally well against a sandy or a rocky seabed.

- **Diet:** Crustaceans, worms, shellfish meat, small fishes
- **Feeding manner:** Bottom-feeder
- **Aquarium compatibility:** Territorial at times

Gobies are found on both sandy and rocky shores. Sand-dwelling species are naturally camouflaged; rock-dwellers can be much more colourful.

Coldwater Invertebrates

Invertebrates may be fantastically beautiful or grotesquely ugly: the sea anemones, starfishes and tubeworms all fit the former description, and some of the shrimps and crabs may qualify for the latter without too much trouble. Among the sea anemones, the Beadlet Anemone (*Actinia equina*) is a very common inhabitant of the coldwater rockpool. *Spirographis* sp. is a fanworm that takes quite kindly to captivity.

Left: **Actinia equina**

Deep crimson Beadlet Anemones are widespread in temperate rockpools. Green and brown specimens are not uncommon. About 2.5cm (1in) high.

Below: **Spirographis sp.**

These beautiful fanworms retract their tentacles after feeding.

Tropical Invertebrates

The majority of invertebrates are even more sensitive to water conditions – and changes in them – than the fishes, and must be acclimatized over a lengthy period (one or two hours at least) when being introduced into the aquarium.

Because many of them are relatively immobile, it is important that their food comes to them, and this necessitates strong water movements in the aquarium. Water currents predominantly from one direction may not be entirely satisfactory, so extra airstones

around the aquarium can provide alternative currents from time to time.

Like many fishes, invertebrates enjoy a growth of algae in the aquarium. They are, however, very intolerant to any metal concentrations in the water, and this includes any of the proprietary aquarium remedies, normally copper- or zinc-based.

The larger specimens can be fed small pieces of shellfish meat or brine shrimps two or three times a week, but the filter-feeders may pose feeding problems. Liquid foods (as available for newly born freshwater fishes) can also be used.

Invertebrates can be kept with fishes (Basslets, Blennies, Boxfishes, Clownfishes, some Damselfishes, Filefishes, Gobies, Jawfishes and Seahorses may be tried as companions, so long as the fishes leave them alone), or they can make up a very interesting and highly colourful species aquarium by themselves.

Right: **Cerianthus membranaceus**
Graceful sea anemones from the Mediterranean and nearby Atlantic.

Below: **Periclimenes yucatanicus**
A dazzling crustacean that would grace any marine aquarium. It is resting among Caulerpa *fronds.*

Index to plants, fishes and invertebrates

Page numbers in **bold** indicate major references, including accompanying photographs. Page numbers in *italics* indicate captions to other illustrations. Less important text entries are shown in normal type.

Further Reading

There is a growing list of books written especially for the marine fishkeeper, but much useful information can also be found in publications covering the wider sphere of fishkeeping in general. The monthly hobby periodicals also carry articles of marine interest, some as a regular feature. Specialist marine fishkeeping societies also publish news magazines, but as their secretaries change almost yearly, it is best to obtain the current addresses from the hobby magazines.

General

Bianchini, F., et al, *Aquaria* K & R Books, 1977
Dal Vasco et al, *Life in the Aquarium* Octopus, 1975
Federation of British Aquatic Societies, *Dictionary of Common/Scientific Names of Marine Fishes* 1982, 1984
    Scientific Names and their Meanings 1980
Hunnam, Milne & Stebbing, *The Living Aquarium* Ward Lock, 1981
Madsen, J. M., *Aquarium Fishes in Colour* Blandford Press, 1975
Midgalski, E. C. & Fichter, G. S., *The Fresh and Salt Water Fish of the World* Octopus, 1977
Mills D., *Illustrated Guide to Aquarium Fishes* Kingfisher Books/Ward Lock, 1981
Norman, J. R., *A History of Fishes* 3rd Edition by Greenwood, P. H., Ernest Benn, 1975
Palmer, J. D., *Biological Clocks in Marine Organisms* Wiley-Interscience, 1974
Sterba, G., *The Aquarist's Encyclopedia* Blandford Press 1983, M.I.T., 1983
Vevers, G., *Pocket Guide to Aquarium Fishes* Mitchell Beazley, 1982
Whitehead, P., *How Fishes Live* Elsevier Phaidon, 1975

Marine Aquariums

Carcasson, R. H., *Coral Reef Fishes* Collins, 1977
Cox, G. F., *Tropical Marine Aquaria* Hamlyn, 1971
George, D. and J., *Marine Life* Harrap, 1979
de Graaf, F., *Marine Aquarium* Pet Library, 1973
Hargreaves, V. C., *The Tropical Marine Aquarium* David & Charles, 1978
Lythgoe, J. and G., *Fishes of the Sea* Blandford Press, 1971
Melzak, M., *The Marine Aquarium Manual* Batsford, 1984
Ravensdale, T., *Coral Fishes* John Gifford, 1967
Spotte, S., *Marine Aquarium Keeping* John Wiley, 1973
Spotte, S., *Seawater Aquariums* John Wiley, 1979
Steene, R. C., *Butterfly and Angelfishes of the World Vols 1 & 2* Mergus, 1977
Straughan, R. P. L., *The Saltwater Aquarium in the Home* Thomas Yoseloff, 1969
Thomson, D. A., Findley, L. T., Kerstitch, A. N., *Reef Fishes of the Sea of Cortez* Wiley-Interscience, 1979
Thresher, R. E., *Reef Fish* John Bartholomew, 1980

Periodicals

The Aquarist and Pondkeeper Buckley Press, Brentford, Middlesex, England
Freshwater & Marine Aquarium R/C Modeler Corporation, Sierra Madre, California, U.S.A.
Practical Fishkeeping E.M.A.P. National Publications, Peterborough, England
Tropical Fish Hobbyist T.F.H. Publications, Neptune City, New Jersey, U.S.A.

Picture Credits

Artists
Copyright of the artwork illustrations on the pages following the artists' names is the property of Salamander Books Ltd.

Clifford and Wendy Meadway: 25, 26, 27, 28, 29, 53

Colin Newman (Linden Artists): 13, 16, 22, 41, 49, 50, 51

Tudor Art Studios: 20-21(B, Ross Wardle), 38

Brian Watson (Linden Artists): 20, 40

Photographs
The publishers wish to thank the following photographers and agencies who have supplied photographs for this book. The photographs have been credited by page number and position on the page: (B) Bottom, (T) Top, (C) Centre, (BL) Bottom left etc.

Heather Angel/Biofotos: Endpapers, title page, 76(B), 104

Aquarian Advisory Service: 54

Bruce Coleman: 79(B, Robert Schroeder), 99(T, Jane Burton)

Eric Crichton © Salamander Books Ltd: 17, 22, 24, 27, 31, 33(C), 34, 36, 37, 38, 39, 41, 42

Jan-Eric Larsson: 14-15, 60, 62, 64-5(B), 68-9(B), 69(T), 72(T), 80(T), 80-81(B), 83(B), 85(T), 86-7(T), 89, 90(T), 90-91(B), 96-7(T), 100(B), 102-3(B), 105, 106-7(B), 107, 110(T)

Dick Mills: Half-title, 44, 55, 64(T), 72(B), 82

Arend van den Nieuwenhuizen: 8, 35, 52, 56, 57, 61, 63, 66, 67(C), 70-1(B), 71(T), 73, 74, 75, 76(T), 77, 78(B), 78-9(T), 81(T), 83(T), 84-5(B), 86(B), 87(B), 92-3(B), 93(T), 94, 95, 96(B), 97(B), 98-9(B), 100-1(T), 108(C,B), 110-11(B)

Mike Shadrack: 108-9(T)

W. A. Tomey: 10-11, 12, 18-19, 23, 32-3(B), 43, 45, 46-7(T), 47(B), 48, 58-9, 67(B), 88, 102-3(T), 112, 113

Editorial assistance
Copy-editing by Maureen Cartwright.

Acknowledgements
The publishers wish to thank The Aquarist (Chelsea) Ltd. for their help with location photography, and Interpet Ltd and Dr. David Ford of Aquarian Laboratories, Halifax, W. Yorkshire, for their help in preparing the book.

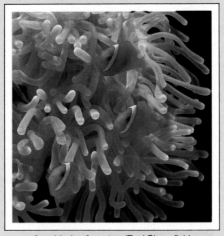

Amphiprion frenatus *(Red Clownfish)*